Got To Tell It

Got To Tell It

Mahalia Jackson, Queen of Gospel

Jules Schwerin

OXFORD UNIVERSITY PRESS
New York   Oxford

Oxford University Press

Oxford   New York   Toronto
Delhi   Bombay   Calcutta   Madras   Karachi
Kuala Lumpur   Singapore   Hong Kong   Tokyo
Nairobi   Dar Es Salaam   Cape Town
Melbourne   Auckland   Madrid

and associated companies in
Berlin   Ibadan

First published in 1992 by Oxford University Press, Inc.,
200 Madison Avenue, New York, New York 10016

First issued as an Oxford University Press paperback, 1994

Oxford is a registered trademark of Oxford University Press

Library of Congress Cataloging-in-Publication Data
Schwerin, Jules Victor.
Got to tell it ; Mahalia Jackson, queen of gospel /
Jules Schwerin.
p.   cm.   Includes bibliographical references.
ISBN 0-19-507144-1
ISBN 0-19-5090500 (pbk.)
1. Jackson, Mahalia, 1911–1972.
2. Gospel musicians—United States—Biography.
I. Title.   ML420.J17S4   1992   782.25—dc20   [B]   91-43947
The publisher gratefully acknowledges permission to reprint
two lines from ''Yet Do I Marvel'' by Countee Cullen. Reprinted
by permission of GRM Associates, Inc., agent for the Estate
of Ida M. Cullen, from the book Color by Countee Cullen;
copyright © 1925 by Harper & Brothers; copyright renewed
1953 by Ida M. Cullen.

2 4 6 8 10 9 7 5 3 1
Printed in the United States of America

To Doris,
poet, partner,
in-house editor,
and the last word

Preface

I admit to what seems a lifelong obsession: A black velvet sound drove me to make four recordings having to do with Mahalia Jackson's voice; a documentary film of her life, with her voice as the score; and now, an impressionistic biography about that voice, where she took it, and where it took her.

My intentions for this book were not to write a formal biography about Mahalia Jackson, but rather to share my personal impressions and reactions to the curve and events of her life as she related them to me, blended with my own observations of her later (both gentle and critical), as her gifts were being acknowledged by a white world with little exposure to gospel music until after World War II (when I and my generation were young).

My account of Mahalia's private and public persona is enriched by four others whose lives crossed hers early and late: her surrogate son, Brother John Sellers, himself a gospel and blues singer; Studs Terkel, her Chicago discoverer; George Avakian, her A&R man at Columbia Records; and Benjamin Hooks, former NAACP director and once her business partner.

The beat and inspiration of gospel music, once heard only in southern and midwestern black churches, revival meetings, and tent shows, is now part of the body and soul of both black *and* white America, together with jazz, the blues, folk, rock and roll, and ragtime.

Now gospel belongs to everyone, thanks, in great part, to the unrelenting, uncompromising energies and pure talent of Mahalia Jackson. Gospel was nurtured by her unique sound and her growth and development in the racial politics of the 1960s; first, when she entered the crucible of commerce, became the star gospel singer of Columbia Records, and made a breakthrough in the cultural and commercial segregation of the recording and media industries. And then, when (at least) the legal battle over segregation was being fought and won in American society, her voice joined that of her hero and beloved mentor Martin Luther King, Jr., and all manner of races and people, to "overcome" in the Civil Rights Movement that made history then, but remains yet the unfinished business of America.

New York   J. S.
January 1992

Contents

Got To Tell It

Yet do I marvel at this curious thing:
To make a poet black, and bid him sing!
Countee Cullen

Let a new earth rise . . .
Let another world be born . . .
Let a bloody peace be written in the sky.
Margaret Walker

All music's got to be "folk" music;
I ain't never heard no horse sing a song.
Louis Armstrong

The Meeting

It was in the summer of 1955 that I found her—Mahalia Jackson of Chicago—at a Sunday afternoon outdoor revival in New York's suburb of White Plains. A group of local Baptist congregations had joined together to present her.

There, in the open-air concert shed, I found myself implanted in a black sea of jubilant parishioners—surely a thousand people from local black churches—gathered for a Jackson "down-home" whirlwind. Under a blue sky and hot sun, the amalgam of her spiritual bewitchery and monumental voice was the most dazzling, gloom-chasing revival Westchester County had ever experienced. Her Sanctified Church style

3

was awesome: Moans and groans and shouts built to shivering climaxes, an exaltation of the life force in a voice that lifted the thousand souls into a state of astonishment, and probably grace.

Except for black church folk, Mahalia Jackson was still an unknown gospel performer for audiences beyond Chicago who hadn't heard her kind of music, much less purchased her early Apollo recordings. And I among them, even though she was, by then, a stellar gospel singer whose rise to celebrity status had come through the good offices of Studs Terkel, the provocative Chicago radio raconteur.

I had never heard or witnessed such a performance before, though weaned on the jazz and blues of some of the greatest black musicians and singers in the New York nightclubs and saloons, from the early thirties to the time of which I write—great black artists like Billie Holliday, Huddie Ledbetter, Charlie Parker, and Louis Armstrong—but never the gospel, never Mahalia Jackson's foot-stomping, hip-swaying, syncopated velocity, hackles-raising affirmations.

In her pink, floor-length organza gown, her black beehive fall piled high atop her head, she swayed, rocked, handclapped, and shouted, pouring it on, snapping her fingers—a monumental body of a woman (she is said to have once weighed over 250 pounds) with her skirt rippling around her legs. Yet she moved light as a feather, a whirlwind with the sensual vitality of an adolescent girl, moving about the stage, between the mike stand and the piano and organ where Mildred Falls, Mahalia's accompanist, reigned with her powerful blues chords, her bounces and triplets perfectly matched to Mahalia's virtuosity. Years later, Mahalia said that she had always visualized herself as a peacock flaunting its feathers all spread out in display.

After the crowd dispersed, I made my way backstage

to meet the artist seated on a bench, fanning herself, her shoes off. She was taking the kudos of her admirers modestly, and, doubtless because I was the sole white face among them, she looked at me as I waited my turn to shake her hand.

After I introduced myself and expressed my awe at the musical experience she had just given me, Mahalia gave me her full attention, and I began to explain why and how I had come to the concert.

I had just completed the photography but not the editing of the first of three documentary films that I intended to merge into one feature-length production. The trio would relate true folk tales of a fading generation of rural America, which had bridged the age of homespun with the age of the atom.

"Cannonsville," the first film of the trilogy to be completed, concerned the threat to the lives and fortunes of a group of farmers of the high Catskill Mountains, whose ancestors had come across from the eastern shores of New England in ox-drawn wagons and made clearings for their log cabins. Now, for the fifth time in the twentieth century, surveyors and engineers from New York City, two hundred miles to the south, had invaded a Catskill Mountains valley (consisting of five villages, among them, Cannonsville), to flood the land, reduce it to the anonymity of a lake bottom, to create yet another water supply for the insatiable metropolis to the south. Families were to be moved off their ancient lands, trees grubbed out by bulldozers, houses and barns burned to cinder, hills levelled; and the bones in the graveyards would be removed to a distant burial ground.

Before I was able to find new investors to help me tackle the two other stories, theater managers screened "Cannonsville," liked it, and began to book it around the country. Those were the days when you went to the movies and got not only a feature film but also a "short subject" to round out the evening program. But as some of the reviews of my film came in, something unimagined and threatening began to happen. Theater managers were regretting their booking, and canceling. Why? Because Pete Seeger, the celebrated folk singer of the day, had written the score with his brother Michael and they performed on the soundtrack of "Cannonsville," and Pete Seeger had now become a notorious member of a blacklisted generation of Americans. He had balked at testifying about his political views before the infamous House Un-American Activities Committee. When Pete was called to the witness stand, he refused to "sing" against friends and colleagues, and receive leniency; he refused to "name names," as that reprehensible business came to be known. But he did propose singing some of his favorite songs for the Committee, in lieu of betraying his friends. The Committee demurred and dismissed him "with prejudice."

That turn of events was the commercial undoing of my little film *then*, because I had no intention of deleting Pete Seeger's name from the list of screen credits. (Pete and Michael had written a haunting and beautiful score based on traditional folk songs.) Booking cancellations were the result. I put the prints of "Cannonsville" deep into a closet, and along with it went months of loving commitment and work, not to mention the gamble of thousands of dollars of personal funds. It hurt *then*, but years later, when the ugly blacklist was expunged from the national scene, the documentary, re-named "Indian Summer," found its way into television, the non-theatrical marketplace, and into the film collection of the Museum of Modern Art. Cold comfort for the 1950s, but a privately felt poetic justice afterward.

Under that hot, White Plains summer sun, with Mahalia still cooling her feet and fanning her large frame, I identified myself as an independent film-maker, and spoke at some length of my plans to add two more stories to the one I had already completed. She agreed to meet me at a New York screening room the following week, with Mildred Falls, her accompanist. The screening of "Indian Summer" couldn't have been more successfully received. The Catskills drama of farmers forced off their land, albeit white, was not an unfamiliar story to the black women. After we shared a meal together, I told them of some ideas that were not yet fully realized in my head: a parallel, real-life drama of black people, their claim on freedom, the end of segregation, and the right to own the land on which they had lived and worked for generations.

Though Mahalia was suspicious of my motives (what did I want from her? what would I pay her?—I learned later that she was always suspicious), the film had moved her. Born and raised on the New Orleans delta, she recognized an all-too-familiar story, though it was as yet an unformed film in my head. She agreed to help me locate a small supporting cast, including a chorus of gospel singers, and she would be the principal performer on the soundtrack. She also agreed that her New Orleans, "back'a town," her part of the city where she grew up and where her family still lived, would be the perfect location choice.

Exhausted but emboldened by Mahalia's promise of collaboration, I left the women at their hotel. Without it, I couldn't have broken through to the black community of New Orleans, and from just the little bit of her personal history that she had begun to share with me, I felt certain it would be a wonderful second film in the trilogy.

The next step would be for me to see my lawyer. The second film seemed to be on its way.

Chicago

A few months later, after the lawyers had made a contract between us, I flew to Chicago to have a getting-acquainted session with Mahalia. In her South Side ranch-style house, a green and gold plush living room, we began a three-day marathon of talk. (As an afterthought that has remained with me for years, I wondered about the extent of her acceptance of me, and whether it was based on my being white or Jewish, or both.)

I placed a small Wollensack recorder on the coffee table. At first I started and stopped it to match Mahalia's initial ambivalence. We agreed that we had to create a chemistry of mutual trust, to dispense with deep-seated cultural bias and walls of racial preconceptions. I began to ask questions, probing about her origins, her reasons for moving north to Chicago, trying not to impose my own prejudices and ignorance of black culture. Ingenuous is not the word. I was trying to be innocent as a child, yet not naïve, or she might meet it with derision. I had to have her confidence, or what I was planning would never work. She had to trust me enough to collaborate; prejudice works both ways.

To her credit, Mahalia helped me get over the initial wariness of a stranger in the house, *almost*. Our meetings immediately became sandwiched between family, friends, and neighbors. Evidently there was always a constant traffic jam in Mahalia's kitchen, all cooking and serving meals, talk and gossip, and not to lose a beat of it, she traipsed back and forth from my question-web and recording machine in the living room—I think, fleeing from it from time to time—to join the hubbub in the kitchen, or take charge of the cooking altogether, leaving me alone, staring at the Wollensack. To her credit, she replied to my queries with increasing candor, one day after the next, but it was being utterly herself to break her concentration and introduce me to soul food, southern-style: fried chicken, chitterlings and greens, pigs' feet, corn bread, and beer. Mahalia invited me into the kitchen to taste the goodies, everyone standing about, bemused, curious about my responses, the white fellow in the living room who was going to make a movie about Mahalia. Food was very important. The excitement in the kitchen was as if it were for holiday preparations; I began to get the feeling that in Mahalia's kitchen every day was a high holiday, which would explain her swaying girth, as ample and commanding as her glorious voice.

It was in Mahalia's kitchen that I met her godson,

Brother John Sellers, for the first time; and, once again, Mildred Falls.

During the first day's session, I handed Mahalia my first check to consummate our deal, a small check for what was going to be a modestly budgeted short film. She stared at it, obviously dissatisfied with the sum. It was in the amount of two hundred dollars, the down payment her lawyer and mine had agreed upon, with subsequent payments later in larger amounts. It was possible, perhaps, that her lawyer might not have briefed her properly, and she anticipated a much larger check to start working with me. I quickly had to assume, by her attitude, that her frame of reference was: A film is a film is a film. A small documentary was the same as making a feature, Hollywood style. My check was a bloody insult.

At first she stared at me in disbelief: I must have made a mistake, left off a zero. She then stared at the check and waved it under my nose. "What do you expect me to do with this? Wipe my ass with it?" She made the gesture. I was shaken by her obscenity.

"That's what your lawyer and mine agreed would be the first payment," I protested. "I would appreciate your depositing it; it puts us into business." She didn't reply, but shoved the check into a drawer of a side table near the couch, then precipitously left the room to join her friends in the kitchen, no doubt to regale them with the story of the bastard in the living room who was trying to get her to work on a movie for nothing!

Curiously, she didn't ask me to leave, and I didn't offer to leave. During the next two days neither of us alluded to the check matter. But on the third day, while Mahalia absented herself for an hour of cooking red beans and rice, corn bread, mustard greens, and ham hocks ("Of course, you'll have lunch with us," said with a twinkle, testing my readiness to partake and enjoy southern-style, black menus, as she had for all the time I was there, standing back to watch the reaction of a visitor in a foreign land), I opened the drawer of the side table. The check lay there, exactly where she had put it three days before. I took it out of its hiding place. When she returned to continue our session, I stood up and held out the check.

"See here, Mahalia," I said aggressively. "You never even bothered to put this into your bank. How can I continue to work with you, if you don't honor our relationship?"

This time she gave me a broad grin. "Hell, man, what am I doin' with you on this little ole picture when Universal Pictures wants me to sign a contract for $10,000 a week? And anyway, they'll drive me around Hollywood in a limousine."

"What film are you talking about?" I asked.

"*Imitation of Life,*" she said, boastfully.

"If you were to ask my advice, Mahalia, I'd say, 'Don't take it.' They want you to play the Louise Beavers role in that film. It's not for you; they want you to play a darky role. Bandanas and rocking chairs and all that Aunt Jemima crap."

Not blinking an eye, she replied: "Man, if you ain't goin' to pay me more than a few hundred dollars to play in your little ole movie, you don' have no business tellin' me to turn down their offer, unless you pay me that kind'a money."

I couldn't tell what she was likely to do next. She walked in and out of the room nervously until she cooled down, and then went back to work with me, as if we'd had no words. I was obviously relieved to have her back on track; it was a risky moment for me. She could have just as easily walked out of the deal. What in hell did she need this crap for? She was going to be in the Big Time in Hollywood.

By the close of that day's taping Mahalia had done a complete turn-around and surprised me. I do think it was the result of my playing back to her the sound of her own talking voice, not singing, which unrolled her growing up years in New Orleans of the 1920s: the little girl, the teenager; the difficult and cruel times; how she harbored little bad feeling for what she had lived through. The more she painted the scenes of her growing-up times, the more I sensed her pleasure in collaborating with me on what *had* to be a labor of love, as much for me as for her.

She even proposed a working title: "Way Back in Those Days". . . .

Despite the disparity in our backgrounds, we had found a way to come together—the vulnerability that comes when private experiences are shared.

I saw the second film emerging, to follow "Indian Summer" in my planned trilogy: New Orleans, stripped of its white tourist trappings; black voices against colonialist servility; the strong black community feeling and solidarity, held together by the glue of the church and the promise of the gospel.

Louisiana Highway

A month later, by pre-arrangement with Mahalia, I flew to New Orleans from New York City, ahead of her, to give me time to get settled in a hotel. She was to follow in a few days, driving south from Chicago, in her lavender Cadillac. Our plan was to drive about New Orleans together. She would bring various groups of people and locations to my attention; we would begin work on the actual pre-production of the film together; her imprimatur would be on it from beginning to end.

Meanwhile, I took modest quarters in a Vieux Carré pension, had my dinners at Commander's Palace, and spent the waiting time wandering down Royal Street, scrutinizing the fashionable shop windows and raffish Bourbon Street.

Mahalia didn't arrive at her brother Johnny Jackson's house on Water Street for another day and a half. I was concerned and called him a few times, to ask if he had any idea why she might be so late in getting into town. Johnny had none of my anxiety.

When Mahalia did reach New Orleans, she phoned me from her brother's, urging me to join them, even though it was late in the evening.

Johnny Jackson's house turned out to be little more than a cabin on a dirt road back'a town. An agitated Mahalia greeted me. She had just had a frightening experience, an abusive, racial confrontation with the Louisiana State Police, and that was the reason for her late arrival. The incident occurred a few miles after she had crossed over the Mississippi at Vicksburg.

Evidently, she had driven all day and was tired, looking for a place to stop, a place she knew would bed down a black traveler. It was dark, about nine o'clock. There was a full moon that lit up the highway bordered by long lines of fences hemming the road on both sides. Open fields and low hills dotted with cabins of black field hands stretched out to the end of the sky, she said. Here and there she passed roadside food stands closed for the night. She was looking for a gas station and a motel that catered to blacks.

Suddenly a police siren ripped the silence. A couple of troopers on motorcycles materialized out of the dark and ordered her to follow them a few yards down the road to a lighted gas station. Parking alongside a gas tank, Mahalia recounted that she smiled at the troopers and tried, in her words, for what I translated as jovial diplomacy. They encircled the Cadillac and ignored her effort at reasonableness.

Though now and then a few cars whizzed by the lighted area, observing a black woman in conversation with a couple of state troopers, she knew no one was likely to stop to satisfy their curiousity. By now Mahalia had gotten out of the car.

She dug into her capacious handbag in search of her license and registration. As she turned the documents over to one of the troopers, she said he put his face close to hers, threatening: "Bitch, tell us why you're drivin' this here car! Ain't yours, for sure." In his head she was sure he was saying: "What nigger woman would be drivin' a lavender Cadillac, if it wasn't stole?"

At this point, Mahalia interjected to her brother and me that she was determined to maneuver herself out of this difficulty; this wasn't the first time she had been a victim of race hatred. Hadn't every Negro in the South been, her entire life?

The police, alone with a black woman at night on a deserted road, were always a threat of serious trouble. She worried that they might know who she was, that it was her practice to carry large amounts of money on her, having always insisted on being paid in cash at the end of every concert—something that, from time to time, *had* gotten her into difficulties. As far as she knew, the police might be in search of money for themselves. She was glad, that this time, she hadn't put any money in her shoes, but was carrying almost fifteen thousand dollars in her oversized bra, and a few hundred dollars in a zippered pocket of her handbag.

The second trooper, not to be outdone by his pal, came up to her with his own question: "Why you out drivin' this car at night? Who's it belong to, woman?"

Mahalia improvised on the spot: "This here's my madam's car. She don' drive, she even makes me have the registration in my own name. Miz Dorsey flew herself down to New Orleans couple days ago, Sir, she had me drivin' it from Chicago to New Orleans, to meet her."

"Where's your Madam live in New Orleans?"

Mahalia, her mouth dry with fear, answered, "Miz Mildred Dorsey, she lives in the Garden District, with her daughter, Mary Lou."

The men circumnavigated the lavender car, enviously. One of them turned back to Mahalia. "Take off your shoes, bitch. You better be tellin' the truth, or you'll find yourself in the lock-up."

She gingerly removed her shoes and stood barefoot on the macadam. It was a cool night, but Mahalia said she was sweating in terror.

"She ain't got no money in her shoes," one of them said. "You got a money belt around your ass, bitch?"

Mahalia said she must have looked real startled, but they seemed to back away from physically mauling her.

"No, Sir," she lied.

"Dump your bag of its contents!" she was ordered.

She upended her bag and when her stuffed wallet fell to the ground, one of the troopers made a grab for it. In an instant, he was waving a fistfull of bills. "I knew the bitch was hidin' cash. Look-a-here! . . ."

They counted out the bills, nearly five hundred dollars, and seemed pleased with their haul. "All right," one of them said, chewing hard on a plug of tobacco. "Let's take her to Jerry Baker, huh?" To Mahalia he said: "Put your shoes on, bitch."

"Who's Mr. Baker, Officer?" Mahalia asked.

"Town Marshall, that's who. Come on, follow us!"

They drove about a quarter of a mile down the road, to a farmhouse surrounded by a high fence. Large chestnut trees shaded the place, so you couldn't tell, Mahalia said, whether the family was up or not. She was ordered to wait in the car, with one of the troopers sitting on his cycle. The other went up to the front door, knocked and was admitted after a wait. In a few minutes he came back and told Mahalia to follow him into the house.

Judge Baker was in his pajamas, robe, and slippers. His hair was all messed, she described, "so he must'a been asleep."

The judge listened intently to the trooper's report and quickly fined Mahalia two hundred and fifty dollars. The charge was "speeding." The trooper holding her wallet gave it back to her and counted out the amount of the fine. He then handed her back fifty dollars. She thanked the judge and the troopers as cordially as she could, considering her rage. "One of those bastards lifted two hundred, somehow, somewhere in between, and they damn well knew who I was! How many nigger-women in Louisiana go around drivin' a lavender Cadillac, for God's sake!" She finished her story, to her brother and me, with a cynical laugh. I winced.

She told us how she had gotten back into the car and how they watched her drive off, how relieved she felt not to have been physically searched and abused: "They would have just as soon put me in the pokey, if they had found all that money I was carrying in my bra and other places. They would have figured, no black woman could make that kind of money honestly . . . would have held me for a week, maybe, ripped me off besides, by the time I could find myself a decent lawyer."

I sat in the shanty house by the river, mute, until her rant and rage against "those white devils" dissipated, as she shared with me what she assumed I couldn't know, white man from the North: that it was not uncommon for southern police to force black women to strip to their skins, to prove they weren't wearing money belts. She had plenty of black friends whose cars' interiors had been ripped apart in a wild police frenzy for hidden sums of "nigger money." She had been lucky this time.

I excused myself and returned to my room in the French Quarter.

It took Mahalia several days to settle in, following the ugly road incident, but after that she began to show me New Orleans, particularly the segregated sections, where we met with ministers and their congregations who would become part of our "crowd scenes."

Segregated New Orleans demeaned me, appalled me, frustrated me from being a good host. I was unable to treat Mahalia to a brilliant dinner at any of the great restaurants of that city; nor would I have been able to have her as a guest in the homes of my white friends who lived in the fashionable Garden District or the French Quarter—if I *did* have white friends in New Orleans. With Mahalia at the wheel of her great lavender Cadillac—due to the city's segregation codes—I had to sit in the back seat of the car. This is what Mahalia told me, and I didn't know whether it was her justified paranoia or the true state of affairs; but I dutifully did as she told me and sat in the back, like the white boss. Because, she said, if a cop came over to check her license, or the car wasn't operating properly, and I wasn't sitting in the back, she'd just jump out of the car and run, and leave it to me. I protested that I didn't drive, but I don't think she believed me.

We drove all over "back'a town," down to the levee, the river front, the docks, the railroad tracks, to the churches, the little private clubs, the school she'd gone to, and to Storyville, the notorious bordello quarter that was no more. One evening we spent in her brother's little house. People lined up on the street and stood there looking at the Cadillac. One by one they came into the house to greet her and welcome her back to the neighborhood. We stood together in the middle of the room as they shook her hand or embraced her. Mahalia would turn, point at me and say, "Here's my white friend from New York, we're gonna make a movie together about the way we live here, baby." They would smile, shake my hand. Later, her brother jokingly told me I was the third white man ever to come into his house. The other two were a policeman and a tax collector.

Wherever we drove, despite the cordial reception given me, because I was with her, it became increasingly clear that black alleged passivity would soon explode into a rebellion, massive enough to push the entire country to its knees, one of these days.

And I write of those 1950s not with hindsight; the signs of the future were in the air.

"Way Back in Those Days"

What much of the world sees in New Orleans is the French Quarter—the old Vieux Carré—and the streetcar named *Desire*. The leafy neighborhoods and squares are ordered, human-sized, perfect for sauntering and dreaming.

Nothing is grander than the city's trademark: the great houses of ancient lineage, faced with lacy wrought-iron balconies and winding staircases; the venerable Cathedral; the stand of historic oaks in Jackson Park with its great green gates; the gardens with their head-tall iron fences; the myriad cafés and splendid restaurants, the garish clubs and dives; shadowy alleyways; a profusion of street musicians.

New Orleans is a venerable madame. In her kitchen folks mix chicory in their coffee, and outside, bury their dead in stone crypts above the marshy ground. Ubiquitous is black music—*aura popularis*—diverse and wondrous. Ragtime, jazz, the blues, gospel.

There is the other, immemorial side of the city between the shoreline of the Mississippi and the old railroad tracks, home for tens of thousands of working-class blacks and Creoles with their racial mix of black and Mediterranean origins: a planless, poor yet homey district of old clapboard cabins and small houses.

Here, between Water and Audubon streets, Mahalia Jackson was born in 1912, the third of six children. She called the house she grew up in a "shotgun shack."

"In those days," Mahalia recalls, "the houses were pretty shabby, and many of them are the same way today. The rent was no more than six or eight dollars a month, and things were pretty lean. As a girl, it was nothing for me to go down by the levee on the waterfront and pick up wood that had drifted off the river onto the banks. I'd let it dry, and then carry it back on my head, for cooking, and to keep us warm in the winter. You could be in the house and you could see the sun outside, through the roof. If it rained, it rained inside. We'd rush about putting pots and pans around the floor to catch the run-off before we got flooded."

Whites owned the grocery stores and bars of the neighborhood. Most of the black women worked as domestics in the houses across the river. The younger men worked the docks, manned the riverboats, fished and crabbed in river coves, and did track work on the railroad.

Her father, like many of the men of the neighborhood, found work on the riverfront, the docks downtown, and on the boats. He worked like a dog, moving bales of cotton by day and doing barbering after supper. On Sunday he preached in a Baptist church. Money was always short, but that was the common experience of blacks. And Mahalia was forever mystified: it was a miracle how they managed to survive. Segregation had maintained poverty and separation, as if the Civil War had never happened.

Her mother died when Mahalia was five. Her father took her and her ten-year-old brother, William, to her Aunt Duke's house nearby, where she lived until she was sixteen. Aunt Duke worked as a cook for a wealthy white family in the Garden District, and she was "a bitch on wheels" who beat Mahalia and her brother with cat-o'-nine-tails when she decided that they weren't "toeing the line" to her satisfaction.

Mahalia made nightly visits to her father's barbershop. He had acquired another family and new children along the way, yet he gave her a great deal of love and enough money to give Aunt Duke for her brother and herself. Mahalia never forgot the heavy work she and her brother did as children. William worked as a yard boy for a number of white families, earning a pittance; Mahalia did household chores for her aunt after school, scrubbed the cypress-wood floors with lye, made mattresses from corn-husks and Spanish moss, and split-bottom cane chairs from sugar-cane and palm fronds. She never got paid for any of it, and never forgave her aunt for the abuse.

Both sets of Mahalia's grandparents were born into slavery. From her Uncle Porter, her mother's older brother, she came to know about the horrors of slavery and the continued enslavement of allegedly free citizens, in the post–Civil War era. As far as her Uncle Porter could figure out, the 13th and 14th amendments to the Constitution had never happened. He told her how the former slaves were defrauded of the tiny wages they earned chopping cotton on the leftover plantations, working twelve hours a day—men, women, and children. It was rare, he told her, for a black person actually to receive the fifty cents a day he earned, because the plantation bosses charged the earnings of field hands against their room and board. How could a man call himself free when he was forever dependent, poor, and politically impotent? However her simple but wise Uncle Porter explained such things to her, it affected Mahalia so powerfully that it shaped her business relationships for a lifetime. In her professional years, whenever she sang for money, she demanded her fee or a share of the "take" even before the close of the concert, and then, *always in cash*. (That clearly explains the fifteen thousand dollars deposited in her ample bra on the Louisiana highway. Her Uncle Porter had taught her to trust no one in this world.)

A small child's face set itself upon her broad, glossy grown-up features whenever her thoughts reached back to the little girl "back'a town." Her brow furrowed with the inconsistencies: On the one hand there were segregation laws, they went to separate schools; but after three o'clock, black and white kids played and battled together, and the battles were without racial prejudice, Mahalia remembered it, with some rue. The racism of the parents hadn't touched the children, strangely, until the time for growing up came, and then custom demanded that bigotry be the nature of life, and anyone who thought differently was a freak. There were occasional ugly confrontations with neighborhood Italian kids, who would beat up on her and her friends for one reason or another. "We always defended ourselves, but we always made up," she had laughed, in recollection. "They're probably all still there where I left them forty years ago."

With special affection Mahalia recalls one white family named Ryder. They lived in the Garden District, in a big, sprawling white house. The Ryders were unusually generous with their black employees, among whom, from time to time, were members of Mahalia's family who worked in the kitchen as cooks, served in the dining room, or were chambermaids upstairs.

It was the Ryders' practice to honor the local custom of "the pan"—the sharing of leftovers from the family's Sunday and holiday meals. The food was set aside in a large pan for the employees to carry home. "The pan" was often the difference between having enough food to tide them over the weekend or a long holiday, or going hungry.

When Mahalia was in the eighth grade, along with doing the chores for Aunt Duke, she found additional work as a laundress, and had to handle an extra five-hour stretch of hard work—five hours of school, five for Aunt Duke, five for herself. It was a heavy load for a young girl, but the last five hours gave her her own money, which she saved religiously.

And then there was her voice. Even as a small child of eight, Mahalia knew she possessed something special—an uncommonly large voice. Everyone praised her for it, and she used it in and out of the church. Music engulfed her; she was music. She acquired a rich repertoire of spirituals and hymns, and the more worldly popular songs she heard on the early phonographs. Music, as well, was all around her. New Orleans was filled with performing bands; pianists worked in the parlors of the bordellos of Storyville; small combos played on the showboats for white customers and black employees while they steamed up and down the Mississippi.

It seemed as though everybody in the City of New Orleans knew how to play an instrument or sing a song. If they didn't, they filled up the saloons to hear somebody else sing or play. In the years before the first World War, anyone who could afford it bought a phonograph with a cranking arm, and collected records.

The new music was everything, produced by the blacks and for the blacks. It became the fashion to hire brass bands to lead fancy black funerals to the graveyard and back. Mahalia explained: "If anyone in the community would die, who was well-loved, or belonged to the secret orders, like the Knights of Pythias, they would always hire a band. To this day, it shows the good-time spirit of the city. To cry at the incoming of a child and rejoice at the outgoing.

"When I was a girl, the coffin would be drawn by two or four white horses, and the Knights would march in their uniforms which made a great spectacular. They would hire a band if it was a man that died. If it was a woman? They didn't usually do it for us. But if a man died, he had a band. And when the body would come out of the church, the drums would hit up sad songs like 'Nearer My God to Thee' or 'What a Friend We Have in Jesus' and they would march behind the hearse with the coffin, all the way back to Green Street cemetery or whatever ones were close to the community.

"After the burial, the band would strike up these religious songs, and the people from all over the city would meet at the cemetery and return, dancing in the streets to 'When the Saints Go Marching In'—that's what that one's all about, baby—or whatever they felt like playing. The children and the old folks would come back from the cemetery, walking or on bicycles, on trucks or wagons. They would all get right into the jubilant feeling of this jazz music. So that's how a lot of our songs that I sing today has that type of beat, because it's my inheritance, things that I've always been doing, born and raised-up and seen, that went on in New Orleans.

"And when I came up north, and a lot of people questioned me about the way I sing these religious songs, I tell them: that's the way I heard them being rendered this way from my childhood. That's why many think it sounds like jazz, but it is the way I heard it played as a little bitty girl. The type of music I'm talking about, known to all New Orleans people after they buried their dead, was 'Second Line.' White folks would stand along the curbs, with the blacks who hadn't gone all the way to the graveyard. The returning band and the dancing crowds always attracted attention. People were always happy then, even though they knew someone had died. But that music did mean something to them.

"That type of music is really my soul. That type of music is just like—what is the great song that the Italian people, and the Irish people love so? Each country's got their own folk song they love. Well, this type of singing, and this type of doing is just me! It's part of New Orleans people, the things they do. It's like eating red beans and rice!"

Mahalia saw little difference between gospel and folk music. She rebuked many critics who implied that her music was too easy, that it didn't take a long study. She had a ready answer for them before she even left New Orleans for Chicago:

"Some people are a little ashamed of gospel songs and folk songs, because it doesn't take a lot of long study, and they *are* simple songs of people's hearts. Sometimes, folks don't even think it is art, or what they call 'art'—something complicated, and you have to go through a long period of study; they think that if a song comes from the heart, then maybe it's just too easy. Well, I don't agree with them!

"No one can hurt the gospel because the gospel is strong, like a two-headed sword is strong. Strong enough to cut through, no matter if Satan himself sings a gospel song. It is good to see everyone try to sing it, because there are some that make a big mockery of it. These songs are really the staff of life for some people. Now, take the blues, they sound very good to me, they are part of our great musical heritage. And I wouldn't say it isn't one kind of our finest songs, but it just doesn't give you much relief.

"You know, it's just like a man who is a drinker, and when he gets all through being drunk, he's still got his troubles around. For me, it all depends upon what condition or frame of mind I'm in. Songs fit my occasion. But gospel songs I like to sing for myself, songs like 'Jesus, Lover of My Soul,' 'My Faith Looks Up to Thee,' and 'Just As I Am,' because—

"Sometimes you feel like you're so far from God, and *then* you know those deep songs have special meaning. They bring back the communication between yourself and God."

Good Times, Bad Days

Young Fred, Mahalia's cousin and Aunt Duke's son, earned a sometime living in music. He brought home the new recordings, and he and Mahalia listened to them for hours on Fred's phonograph, late into the night. That's when Mahalia first heard Bessie Smith sing "Careless Love," and it was to haunt her all her life, even though Mahalia eschewed "the blues" for herself.

Cousin Fred, though still in his teens, tried to live the sporting life in New Orleans. His strides, though, were too small to take him any distance. Fred was no saint, but he wasn't much of a sinner either, was the way Mahalia put it.

He was her confidant, her pal. It was Fred who urged her to make plans to escape the clutches of his mother, Mahalia's nemesis.

Fred played the new records for her and made her yearn to try out her voice in the world. He also fed her own restlessness, her wanting to escape New Orleans for the exotic world up north. But Aunt Duke had her on a very tight rein. Mahalia became the secretary for many social clubs Aunt Duke thought important, and it was also a means for her aunt of knowing everybody's business.

One day, Mahalia's father went to Chicago on a scouting trip, but he didn't like the bigness; not the least of his concerns for his family was his fear of Al Capone and all the other resident gangsters. No, it was New Orleans for cotton hauler-preacher-barber Jackson, and for a time his negative report on Chicago frightened his fifteen-year-old daughter.

Beyond her own family, the earliest influence in her girlhood was what went on in the Mount Moriah Baptist Church. If you helped with clean-up and scrubbing, the preacher gave you the gift of ringing the bells for early morning service. On Saturday night, the church would screen silent movies for an admission fee of only a nickel or a dime, and that was much cheaper by far than the price of the New Orleans segregated movie palaces. Her greatest pleasure as a child was the evening and Sunday morning services when the sinners would rise from their seats, come forward to stand in front of the preacher, and get saved. On Baptist Sundays, many of the women dressed themselves in white robes and led the whole congregation out of the church and down the dirt road to the levee, singing, "Let's Go Down to the River Jordan." Reaching the riverbank, the preacher would walk slowly into the river followed by the whole congregation. He blessed the waters and baptized everyone.

Years later, she was still telling listeners that it was the foot-tapping and hand-clapping of the congregation in the Mount Moriah Baptist Church that gave her *the bounce!* "I liked to sing the songs which testify to the glory of the Lord. As David said in the Bible: 'Make a joyous noise unto the Lord!'

"That's me!" Mahalia said.

Though she was to be a Baptist always, it was the Sanctified or Holiness church that affected her life and art. She adored its music. The Baptists sang songs and had an organ, but the Sanctified church used instruments as mentioned in the 150th Psalm of David: cymbals, drums, strings, and tambourines, accompanied by vigorous hand-clapping.

> Praise Him with the sound of the trumpet
> Praise Him with psaltery and harp
> Praise Him with timbrel and dance
> Praise Him with stringed instruments and organs

Mahalia spoke with a wild affection for the Sanctifieds: "Everybody sang and clapped and stomped their feet, sang with their whole bodies! They had the beat, a powerful beat, a rhythm we held onto from slavery days, and their music was so strong and expressive, it used to bring the tears to my eyes."

The Sanctified church members call one another "saints" and believe their doctrine of sanctification has blessed them with the power to speak in unknown tongues, to heal and prophesy. During church services, rites and ceremonials occur for exorcisms and evocations to chase away the evil spirit of Satan. Medicines and medical aid are refused because the healing of the Holy Ghost, the spirit of Jesus, is not human. No human physician, they believe, can heal them.

Another aspect of the Sanctified church next door that drew the young girl to it was its custom, known from the Bible, of "speaking in tongues." It was as if the brothers and sisters had mysterious spiritual messages from beyond, that they wished to bestow on their neighbors an act of grace. For non-believers, their actions were incomprehensible, but for Mahalia, the girl, it was the emotional substance that was to sustain her, creatively, all her life. She believed in "the power of tongues," and she knew *she* had the Holy Ghost. She was "carried away by the spirit." By that she meant: the extraordinary passion that filled her, the prodigy of musical talent that possessed her as she performed, transported her, literally, out of herself. Mahalia could be unutterably happy.

The music of the Sanctified church was the leaven that raised her spirit, especially the "shake-rattle-and-roll" services. Though Baptists were family and friends, she felt them too severe for her taste. A surge of exultation was what she needed to sing the gospel. The pleasure she got from that feast was sybaritic. She needed the intoxication of the Sanctifieds to help her serve her music and her God, to embrace her congregations with love.

Fred had gotten an offer of a music job with a small Kansas City band. Mahalia was happy for him, but shortly after his arrival in the Midwest, he found himself in a nasty jam and was killed in an after-hours saloon brawl. He came home in a coffin. For Mahalia, Fred's death was the signal that it was time for her to make plans, move out of Aunt Duke's house and out of adolescence.

With Fred's death (and the departure of many of the other younger black men to take jobs as porters, chefs, and waiters on the big, new, shiny transcontinental trains) there was little left in New Orleans to hold back Mahalia from trying her wings.

"You may not be aware of it, but New Orleans is the type of city where people believe in living today and be happy. It's always been a very merry city where people enjoy life to its fullness."

By the time I had met with Mahalia in Chicago, in the mid-1950s, she had reached a mature private plateau where the calling up of childhood memories in distressful years could be done with astonishingly little bitterness. (The accumulation of nearly a million dollars by then, in cash and property, probably didn't hurt.)

One day, she seemed especially gluttonous and eager to taste her "merry city" roots again, her tenses swinging dizzily from past to present to past. What had been was still Mahalia.

"As poor as most of the people are, they'd have these good-time days. It's an old New Orleans custom, that when they've finished work on Friday, they start good-timing from Friday till Blue Monday. New Orleans is noted for Blue Monday. Folks wouldn't work on Monday, they'd just have this good time," she chortled. "When I was growing up there, a lot of people didn't worry about getting rich. They didn't even think about making much money; but they didn't have nothin' . . . they still had this wonderful spirit under segregation, because my people stayed by themselves, creating their own fun. Sure, they had plenty of tragedies, everybody did, but somehow everybody went along, got along.

"Now, Fred and me used to listen to the Dixieland music at the various dance-halls in our part of town. Then, we just called it 'music.' Folks later came to call it 'jazz.' And that type of music was what the better-thinking people called 'indecent music.' It was the music played for common people, you know, in the honky-tonks, the saloons—music that the world raves about today. I knew about jazz music when I was just a girl. King Oliver played on the trucks in our part of town, or when there was going to be a dance at the Bull's Club downtown, or at the Pride-of-Carleton.

"But there were many things that went on in my hometown that I didn't like. I never did like the world-famous Mardi Gras that went on in New Orleans. It was a beautiful sight, but to me it was horrible. I have seen so many people hurt on that particular day. What went on would turn it horrible.

"I can't forget how the poor people would spend all of their earnings, from one year to the next, so that they could dress up like a tribe of beautiful Indians, or like the head of the Zulu tribe in Africa. A lot of women would be dressed in expensive clothes, and all this dressing-up, the beautiful way they looked, many would never get back home. They could get killed that day in New Orleans.

"People were free to cover their faces, and any crime that they committed, there were never any charges made against them.

"Yes, many dressed up to look like different Indian tribes, and they would buck each other that particular day. They called themselves the Red, White, and Blue, or the Yellow Pocahontases, and many others I've forgotten. In the 12th Ward of New Orleans, men called themselves chiefs of certain tribes. Folks danced and jumped about, sang and embraced each other. That was Mardi Gras 'back'a town.' And this would go on for a whole week!

"The white people would celebrate *their* Mardi Gras with big and expensive floats that went down the main part of Canal Street, which were very beautiful and high class. Every-

body came from different parts of my section to see the
Queen when she came in on one of those beautiful floats that
cost millions of dollars. People came from all over the world
to see this.

"But for *my* people, for them it would be such a tragedy.
If one of the tribes demanded that another 'take low,' you
know, bow to them, they'd kill each other and nobody was
punished! The State, the law never did anything about the
killings." (Who cares if a black man kills another black man?
was still the challenge in her flashing eyes.) "I hated that
condition so much," she went on. "Can you believe it, people
were free to shoot and kill each other?"

Then, there was her residual anger that she couldn't shake off,
at the scenes of violence having to do not with good-time-
gone-crazy but with the way civic relationships were *always*
gone-crazy.

"There was a very bad police system back then," she said
quietly. "They'd run my people in, if they just saw them
standing about in their own community. Many were put in
jail for no reason, got beaten up, and some of them were
killed. Resisting an arresting officer was enough to get you
into serious trouble. It made us feel we never had any police
protection or justice.

"There were no black lawyers that I ever knew in New
Orleans. If a black man went to jail, and he wasn't thought
well of by certain white people, he could be treated pretty
bad, even manhandled by the police. A man was a man; and
some of the blacks resented the police for mistreating them so
bad.

"If my people saw the police coming up anywhere,
they'd scatter like flies in a pool-room; they'd run like some

King Kong was coming. Those things made for a bad environment, and made our friends and neighbors very sad.

"The preacher, at that time, was the only mouthpiece that my people had. He would go down to the courthouse and ask the judge to show some mercy to his people. If a girl went astray and became a 'bad girl' who was working in one of the houses in the red-light district, her mother always went to the preacher for help, to get her away from the house of prostitution and come back home. And if a boy got so bad he'd become a wayward boy, his family would ask the preacher to see the judge. The people respected the old preachers and had faith they could save their children, get them home, or pardoned. Even if they'd committed a serious crime and went to the penitentiary, the preachers could get the judge to soften their punishment, because they were good people and were listened to.

"I always loved the church because of its powerful music. When the people would sing, I always did love the way the congregation would sing a song. It seemed that it had a different tone quality than the choir would have. I would always find myself drawn to the church. It's because I like the songs and I like the way the preacher, the old preacher would preach his message. He weren't educated like some of our ministers today, but there was a way that he would preach, would have a singing tone in his voice, that was sad. And it done something to me.

"It *is* the basic way that I sing today, from hearing the way the preacher would sort of sing in a—I mean, would preach in a cry, in a moan, would shout sort of, like in a chant way—a groaning sound which would penetrate to my heart."

Heading
North

By the fall of 1927, Mahalia overcame her fear of going north, despite her father's anxiety about "Chicago gangsters," for she had a passionate trust in the church, and was sure it would protect her from the evils of the big city. Her Aunt Hannah, who lived on the South Side of Chicago, had come to New Orleans for her annual family visit. There were rumors that Mahalia wasn't the first of the children to be invited by Aunt Hannah to come and live with her, but Mahalia was the one who did go; even as an adolescent she was a sharp and persuasive negotiator.

Young Fred, gone to his early grave, still marked her

with his persuasive urging: she must seek her musical destiny in a fresh, new place. She knew that she must and that she would prevail, away from the clutches of his mother, Aunt Duke.

Mahalia and Aunt Hannah boarded the Illinois Central with heavy, rope-tied suitcases and a large food basket, found two hard seats of worn green plush in the segregated coach directly behind the engine, and tried to make themselves comfortable for the three-day trip to Chicago.

Sixty years ago, interstate accommodations for black travelers, by custom and law, were required to be "separate but equal," according to an 1893 decision of the Supreme Court, but blacks who may have purchased first-class tickets for train travel very often ended up in the baggage car if segregated coach seats were all filled. The women curled up close to each other at night, covered by a woolen throw Aunt Hannah had packed, in anticipation of the unheated car. The dining room was off-limits to blacks, though, ironically, all the snappily uniformed waiters recently employed by the railroad were young blacks; so Mahalia and Hannah ate from their food basket filled with sandwiches, fruit, homemade pie, and containers of fresh water and milk. At station stops they were able to get hot coffee from the peddlers, who did a lively business through the open window servicing black passengers.

The promise of good fortune waiting in Chicago made the uncomfortable three-day trip bearable for Mahalia, but for the rest of her days, whenever she traveled by train, the initial racist trauma of that first trip never faded from memory, even when she could afford private, first-class accommodations.

Young Mahalia was traveling with almost a hundred dollars pinned to her bra, a practice that became a lifetime habit.

It was the money she had scrimped and saved, working as a laundress, a nursemaid, running errands—the extent of her capital. She stared out the window at the dawn light as the train emerged from the prairie darkness to confront the Chicago skyline in the morning light. Despite not sleeping for three days and sore muscles from having to scrunch up on the worn, hard seats, the vision of the metropolis was thrilling, and she knew she was arriving where young Fred had told her she was supposed to go.

Life in Chicago's South Side wasn't much different for Mahalia from what it had been in New Orleans; she had to find what she called "wash jobs" in laundries or in the white homes of the rich North Side—hard and dirty work—while she dreamed of becoming a nurse. But Chicago was where a black could find an open door somehow, and one opened for her when her aunts took her to the Greater Salem Baptist Church; she quickly became a member of its choir. Whatever misery the outside world dealt her, there was always music and the church.

The Great Depression of 1929 arrived like thunder and struck down the already poor. People were sleeping in doorways and city parks, picking scraps of food out of garbage cans, standing in lines for bread or a bowl of hot soup. The banks crashed, and Franklin Delano Roosevelt entered the White House armed with the National Recovery Act—a series of laws to put people back to work on the farms, in industry, and in the rebuilding of the cities.

The black population suffered acutely in the Depression storm, and Mahalia, as a witness, noted ironically that those who had stayed in the South seemed to suffer less than those who had emigrated to the North. Conditions had always been on the knife-edge of despair down south, but at least you could plant a few greens and find fish in the rivers and lakes to put on the table. The Depression was just more of the same for southern black folk; there was little to distinguish between the poverty of infinite duration and that of a national catastrophe.

Much younger than Mahalia, Dick Gregory, the black comedian who had also lived in Chicago those years, has wrapped up his view of the Depression era: When he was growing up, they were so poor that he and his siblings had to take turns going to school—there weren't enough shoes or clothes to go around, and they were always hungry. His mother took in laundry; when she hung out the sheets to dry on the windows, it was the only time they had curtains. And *then* came the Depression!

Chicago's South Side, like every minority neighborhood in America, had to be creative to survive. If legitimate business and jobs were non-existent, folks had to create new ones to keep alive: gambling dens, speakeasies, store fronts to sell policy number games, "buffet flats"—the blinds for bordellos. Those were terrible times for Mahalia, but she was sure that the Lord had His arms around her: "The Depression was responsible for my whole career in gospel singing," she said, thinking back to those desperate days of her girlhood.

She became a member of the Johnson Singers, a group that sang in the neighborhood churches for as little as $1.50 a night. The money came from the pass-around plate, the only way churches could pay for coal heat and monthly mortgage charges. In time though, the Johnson Singers became inundated with invitations to sing in churches downstate and in Indiana; soon they performed as headliners for the out-of-town Baptist conventions. But Mahalia managed to save only a few dollars for herself, after turning over a small contribution to her aunt for the rent.

The only music lesson she ever had took place in a South Side music school run by a Professor DuBois, a tenor of local fame. After listening to her sing a few songs, he remarked that she had better stop hollering, and *then* she ought to change her style of singing; it was no credit to "the negro race"! Professor DuBois urged her to sing slower and sweeter; in this way, he insisted, she would appeal to more white folks who could better understand the songs she was singing. That ended her one and only music lesson.

Gospel singing in the early thirties was parochial. It was, for the northern blacks, the music they had left behind in the southern churches, it was mother's milk longed for, hungered for. Mahalia, with her mesmerizing, special harkening of the singing gospel, was passed along on a kind of circuit, so for pennies a night the churches got to know the rising star. She went as far away as Buffalo, where she appeared with a trio, and the audience paid a nickel for admission. Life was hard, but Mahalia was sure that God was saying, "Let there be light for Mahalia." On those church tours, she would stay the night at the minister's house. After the service and her performance, he would feed her, and they would sit in the kitchen divvying up the night's take, less her room and board.

In between the church dates, she looked for work in Chicago, but pointedly avoided factory jobs, unless she didn't have a penny in her pocket. She hated the violent way of life imposed on girls by the system—the stress of the speed-up on the production line, or the private feuds of the girls, nasty confrontations often put down with force by the supervisors. Many of the girls would need emergency treatment because of blackened eyes and broken bones. When things went well, the girls took home sixteen or seventeen dollars a week, if they were lucky. Work as a chambermaid was what Mahalia had known her entire life, and it was what she looked for—a twelve-dollar-a-week pittance in a city of no jobs, but it was better than being turned into an assembly line freak, like so many girls of her age. Even if it was bare subsistence working in a hotel or rooming house, combined with the church week-end trips singing the gospel, she knew she was better off than most girls her own age. But that didn't allay her anger about how almost impossible it was to survive, how depressed she could be if it weren't for her singing of the gospel that lifted the heart.

By 1935, Mahalia was twenty-three and had lived in Chicago for eight years. She spoke of that self as still being a "fish and bread singer." She sang for her supper as well as for God.

It was at one of the many church-sponsored socials that she met her first serious love, Isaac Hockenhull. He had been educated at Fisk and Tuskegee universities, specializing in chemistry, but after graduation there were no teaching jobs for chemists, much less work in Chicago laboratories, so Ike became a mail carrier in the Chicago post office. He quickly became a central figure in Mahalia's circle, a serious young man with a resonant voice. Ike was ten years older than Mahalia, and she wondered why such an older man, one so well-educated, would find her of interest. But Ike was as serious about her as he was about most everything else in his life, she decided. He believed that she had a glorious voice, one that could carry her into concertizing. Soon, nothing mattered to him more than marrying her first, then seeing that she got her voice trained to be a concert artist.

Ike was always in the audience when she sang anywhere in Chicago, in those early days when she was still with the Johnson Singers and beginning to receive serious attention in newspaper reviews. Wildly in love with her and her musical genius, Ike finally convinced her that living together, combining whatever money they had, made a great deal of sense. They married in 1938.

At about the time Mahalia met Ike, she had decided to go into business for herself, to do professionally what she did for her friends for free—hairdressing, even though the competition to become a hairdresser on the South Side was enormous. Black women with few talents at their command found hairdressing an easy craft. Many worked out of their apartments; others turned empty storefronts into salons. An upright church-going woman could rent an empty store front for twenty-five dollars a month. If one didn't have the ready cash, an obliging landlord could be found who would extend three, even six months of credit, for vast stretches of South Side real estate were going begging.

Ike's mother had been in the house-to-house cosmetic business in St. Louis, and made a good living by producing her own brand of lotions, creams, and oils and selling them in the street marketplace under the name of "Madame Walker." Ike, the trained chemist, developed his own formulas for his mother at the kitchen stove late into the night. When he married Mahalia he encouraged her to add cosmetics to her hairdressing business. While making his postman's rounds, he would sell the items they were producing at home to people on his route. And Mahalia, when she made her short singing tours out-of-town, would pack a small valise with samples of their cosmetics and lotions. They were doing well together. By 1938, Mahalia had achieved a modest reknown locally, with her singing and church work. With her hairdressing, she developed quite a clientele and called her beauty parlor "Mahalia's Beauty Salon." Soon there were five girls working for her. Then she figured that she could expand the business by adding a floral shop to serve the churches with floral displays, especially at funerals for which she sang. There were mourners who insisted that she sing at their funerals, or they would refuse to buy flowers from her—gentle blackmail to get her beautiful voice into the service. The flower business turned out to be very profitable.

All of this activity propelled her right into the local commercial scene—from singing in church tent shows, down-at-the-heel storefront churches, to large ballrooms. She would begin the evening by selling tickets in the box office or at the church tables, discovering that she was good at chit-chat with people as they arrived, as well as keeping a close eye on the till or cigar box. Becoming sophisticated in the ways of show business was what she was learning: "Gospel songs had become so popular that men who didn't care about the church or religion were moving in to prey on the public and the singers. I began to have real trouble with them on the road . . . and some were up to such slick tricks with the cash box, I had to make it a rule *not* to sing until they handed me the money they promised me, *before* I sang." Those experiences triggered in her a demand that she get paid in cash every time she made an appearance.

She learned quickly not to be "sub-contracted," as she put it, by any promoter. She had come a long way from the days of sitting at a minister's kitchen table and divvying up the church's pass-around plate.

Ike was convinced that his wife had the greatest voice in America. He spent a good deal of time thinking about how to go about getting her into "the big time." He urged Mahalia, who had never had a voice lesson, to meet with Madame Anita Patty Brown, a celebrated voice coach of the South Side, who once had been an opera singer.

Reluctantly, Mahalia agreed to meet the lady, who quickly confirmed Ike's conviction that his wife, indeed, had a singular voice. He was now convinced that Mahalia must get into the world of the concert artist. Mahalia couldn't help but be secretly pleased with the Madame's evaluation of her voice, but was absolutely opposed to the idea that she should be singing classical arias from operas and operettas. For Mahalia, it must be gospel, whether Ike liked it or not. All of the attempts by Ike—the college graduate with cultural notions totally alien to hers—to polish what he thought was his rough diamond failed. No argument he came up with could move her from her determination to break through with the gospel on her own terms. Ike had to understand that the gospel was her music to the bone, and no one would dissuade her from feeling that there was any singer alive who could match her vocal power when she sang who and what she was.

At the same time that she received Madame Brown's enthusiastic endorsement, Mahalia had begun to work with Thomas A. Dorsey, the leading gospel composer and coach of the day. He was the choir master for many of the Baptist churches in Chicago, and in the twenties Dorsey, under the name of Georgia Tom, had gained a reputation as composer-pianist for the great blues singer Ma Rainey. The two had traveled all over the country playing his celebrated gospels: "Precious Lord," "Walk All Over God's Heaven," "I'm Goin' to Live the Life I Sing About in My Song."

When Mahalia and Dorsey met at chorus rehearsals, he assessed her talent as far above average and aimed at becoming her coach and mentor. He soon realized that she was a natural-born singer who needed no outside direction, nor did she have the disposition to tolerate anyone's advice. He settled for introducing her and his songs to the Chicago and out-of-town churches, where they were always ready to embrace a new gospel voice. The fees were small, but Mahalia didn't care as long as the publicity was substantial. Cannily, she figured that the fees would grow along with her reputation. By the time Mahalia became a star, Dorsey had dedicated a number of gospel songs to her and they became her trademark.

Someone once asked Thomas Dorsey about the meaning of gospel, and he replied: "Nothing so prepares the heart and mind for the reception of God's word as gospel songs, spirituals and Baptist shouts, sung with spirit and with understanding.

"Gospel songs bring joy, peace and happiness. They edify men and glorify God. They find their place in the hearts of men. They come into being by the special divine impress, and not as ordinary songs."

As time went on, Ike found it increasingly difficult to keep his mouth closed about his feelings and his wife's ambitions. He complained that gospel was not his idea of "art"; he argued endlessly that she was wasting her talents. Mahalia would counter that she would do as she liked, not what pleased him. Gospel, she would remind him angrily, was *her* life, not his! If he loved her, he would trust her to choose what was right for her. If not . . . It had come to a serious impasse, and created stress in the marriage. She would remind him that, if she listened to him, she should sing opera. If she listened to other so-called well-wishers, she should sing the blues! She didn't need advice from *anyone*, about what they were ignorant of. "Tell me, Ike," she said, "what black person you know couldn't become a decent blues singer if he or she wanted to? I could be the best there is, if I wanted to sing the blues, but I don't! I'm here to sing the gospel!"

To compound the stress, Ike was laid off at the post office, at a time when they happened to be down to just a few dollars. But Ike had a solution. He had heard that there was a Federal Theater project in the Loop, and they were casting for a production called "The Hot Mikado"; it was a swing version of Gilbert and Sullivan's operetta, and they were looking for lead singers. He urged Mahalia to go downtown at once and try out. "There's no one better than you!" Mahalia balked, but saw that nothing would satisfy him unless she tried her luck, and they did need money to come in, some way, some how. She finally agreed, got herself all gussied up, and took the El to the Loop.

At the audition hall, she was told she needed a lead sheet of the song she intended to sing. She wandered out into the streets for an hour or so, looking for a store that sold sheet music. It was all very upsetting and strange: she had never used music sheets, she just sang, and anyone who accompanied her just knew what to play. She found a music store and picked out the music for a song she had sung her whole life: "Sometimes I Feel Like a Motherless Child."

When it was her turn to sing, back at the audition hall, she sang the song as she knew it, without reference to the printed arrangement she had just bought and the pianist was playing. For a moment, it threw the listeners into confusion, not to mention the pianist, but their faces were registering that they were hearing one of the most remarkable of voices. Mahalia was oblivious to their reaction. She sang, excused herself, and left, sure that she had failed—maybe it was her interpretation, or her bravura style. Whatever it was, she felt depressed and wandered about in the Loop for hours, to get up her courage to go home and tell Ike that she had failed.

Ike had come home early with good news. He had found a new job selling insurance. The telephone that was about to be turned off for non-payment was ringing. It was the stage manager asking for Mahalia, and when told that she hadn't returned yet, he left the message that she had won one of the leading roles and they wanted her at the theater the next day, to begin rehearsals.

A euphoric Ike greeted her with, "You won! You won!"

Mahalia sat down to collect herself, while he told her his good news. They'd have money in the house all around!

She looked at him and said: "You got a job today? Then that settles it. I'm not going to sing in that show, I'm not going to any rehearsal."

Ike reacted as if lightning had struck him. He was speechless when she called the theater and said she wouldn't take the job.

Mahalia was furious with Ike for expecting her to add yet another job to her long list of hairdressing, floral design, singing solo and in choirs. How come they had no money? Where was the money she brought home? Until that moment she hadn't allowed herself to face what had been happening to them. Until that moment she had tolerated his addiction to playing the races with their tiny savings. He had begun with small bets, and it had grown into serious handicapping; with Ike always protesting that, in reality, gambling was a business too.

Her rejection of "The Hot Mikado" was the turning point in their tempestuous marriage, it would appear. "We came apart over gospel singing," she would say later. Not to mention Ike's studying the racing forms every day, betting even when he didn't have a job. And there was a time when she discovered that the cash she thought was safe under their bedroom rug had actually been used by Ike to buy a race horse! "There it went, our savings up the flue." It had never been her intention to go into the horse-racing business, she commented ruefully, years after.

Ike and Mahalia separated when it was obvious to her that nothing would dissuade him from gambling away their money. They divorced, but remained friends for the rest of his life. Ike, the charmer, would appear from time to time, to borrow money, and Mahalia would comply with a check for his continued addiction.

Though she and Ike had no children, there *was* a child who came into her life during their marriage. Of the coterie that adored and served Mahalia in the early Chicago years of poverty and struggle, the sole survivor is Brother John Sellers, the godson and protégé she found in a Mississippi church minstrel show.

Little Brother John

His father was a railroad man who had no interest in home life, nor did his mother. John Sellers was born in Clarksdale, Mississippi, in 1924. His mother moved to Leland, a small town ten miles east of Greenville, on the Delta, and when John was five she left him with a mulatto woman, Beatrice Newall, who was famous for operating a sporting house in nearby Greenville with her partner, E. E. Kersh. Their bordello served white men with black women, and light-skinned girls were very often passed off as being white.

When John was no more than six, he was already put to work doing odd jobs in the bordello, waiting on the male customers and girls, bringing them beer and whiskey from the bar, helping to make up the beds and collecting the linens. "My mother left me in the house because she didn't want to be bothered raising me. She didn't care no way that I was living in a whorehouse. As for my father, he didn't know what was going on in that house, or didn't care," said John, painfully, of his boyhood.

Clarksdale, the place of his birth, was drenched in blues music history: the highway where Bessie Smith died in an automobile accident was nearby; it was also where Alan Lomax, the folk-music curator of the Smithsonian, discovered Muddy Waters working as a field hand and recorded his songs. And the little boy John grew up there to be the internationally known blues and gospel singer Brother John Sellers, adding his own name to the history. His songs would testify to those harsh years of growing up lonely, poor, and abused by both black and white folk.

"If you've never been hungry, you can't ever know how terrible it is," he would say, thinking back. "I hate thinking about my childhood . . . try not to go into the deepest part of myself, it's so sad, and I can't forget, fifty years later, what I saw and heard when I was a child. . . . The KKK hanging black men on telephone poles! Burning a man alive with an acetylene torch. I saw it! You don't forget such sights. All the black people had to be off the streets at night. We never had much money when the family was still together, so when I was tiny, I'd go fishing for our supper sometimes. I used to eat the wild greens that grew around the ponds. They smell like pepper grass, and I can still smell them today."

Before his parents went their separate ways and abandoned him to Miss Bea in her bordello, John, the little boy, used to wander the streets, at loose ends, looking for something to do to earn money. He'd loiter on the street corners whenever minstrel men like Robert Johnson and Blind Lemon Jefferson sang the blues for handouts. Black, church-going people called those minstrels "sinners," Brother John recalls, but watching them and hanging on to their every musical gesture and sound tied him irrevocably to the blues for the rest of his life.

Fifty years ago, church tent shows were the theater of the Mississippi Delta. When Brother John was only five, he sang, danced, and played the tambourine in these shows put on by the Sanctified church, its parishioners sometimes called "Holy Rollers." The shows built membership, raised extra cash for the impoverished churches, and attracted both the faithful and the unaffiliated looking for entertainment. The few dollars that the little boy would collect from the audience delighted with his performance were often the only money his family had for food that day.

The summer of 1933, Mahalia had an opportunity to visit with her family in New Orleans, because she had been hired to perform in a tent show in Greenville. Her reputation out of Chicago was still modest, but it had begun to attract the attention of southern church groups, who would pay her fare and a small stipend to sing once a night for four or five days.

There in Greenville she came upon John, now nine years old, who was also performing in the tent show. She was so impressed with his energetic and self-galvanized talent, she got him to talk about himself and his family, and what he told her shocked her to the bone. He remembers her saying: "I just can't understand how your mother can let you live with those people running a sportin' house. And I can't understand how that woman, Bea Newall, lives with that white man, anyway. I want to talk to you, boy, about yourself and your future. A child can't live the way you do."

Evidently Mahalia then did some checking around, because the next thing John knew, she was saying: "I have a surprise for you. I know your Aunt Carrie Spellman. She lives in Chicago near me and goes to my church, the Ebenezer Baptist." He looked up at her, mystified: "You do?" "Yes, boy, I do! And I'm going to tell her that your family has left you with people they shouldn't have, and it isn't right."

Brother John recalls that he jiggled the change in his pants pocket and wandered away from her without answering.

A year after Mahalia had returned to Chicago, his Aunt Carrie came down to Greenville and was furious as she watched him working in and around the bordello of Miss Bea. The strange lady identified herself as his aunt. "Don't you want to come and live with me in Chicago?" she asked. John responded with, Why should he want to? He didn't know her.

"You'll like it up there, you can go to school, you'll meet your other aunts and uncles, boy. I'll talk to Miss Bea about it."

Life for young black boys in Mississippi had molded his character; it was more than understandable why he found it hard to go into his deep self: "Who cared how a ten-year-old black kid lived, or would bother to meddle in the business of that boy? If anybody thought it strange that I wasn't in school (though I did get a year or two in a Catholic school), or stranger still, that I worked full-time in a whorehouse, I'd tell about it to Miss Bea or Mr. Kersh, and they'd ask me, 'Who was the nigger that is messin' with you, boy?' If I told them, gave them a name, they'd go to the house and warn: 'I heard you've been messin' with my boy!' If the person would answer, 'We was just talkin' to him,' Miss Bea would say: 'Just don't talk to him at all. Leave him alone.' In those days, most people didn't care if a black child stayed in school or not, or what he did to earn some money.

"A few days later, my aunt told me that Miss Bea wasn't going to let me go up North with her. 'Oh, no, he don't' Miss Bea said. 'He belongs to us!'

"Then, my Aunt said she was going to take me to see my mother in Leland before she herself took the train back to Chicago. Miss Bea didn't mind, because she didn't have any idea what was in my aunt's head. She was taking me to say goodbye to my mother was what was she was doing. I hadn't seen my mother for a long time; she was like a stranger too. In my mother's house, Aunt Carrie announced that she was leaving with me on an earlier train than Miss Bea expected her to. We were taking the train that night. 'You're not going back to Bea and that man no more,' she said. I burst into tears. After all, she was a strange woman to me. While I was crying, she reminded me that Bea and Kersh were not my kinfolk. 'I'm your kin, your aunt, and you're leaving with me tonight, and that's that!' she said.

"And, sure enough, when six o'clock came, I was on that train.

"A few weeks later, Miss Bea came to the South Side of Chicago looking for me. She was very angry with my aunt for whisking me away from Greenville. But after a few days of arguing, she gave up trying to get me back and returned to Greenville alone. As for me? I wasn't sure I wanted to stay in Chicago: it was an unfamiliar place; I wasn't used to big cities; and I wasn't used to my new uncles. You could say—I'd never been around that close to many colored people and how they lived and acted. Miss Bea's sporting house and servicing white folk was pretty much all I knew. . . . Everything was strange to me. Just walkin' about in those Chicago streets made me homesick for Greenville; I just couldn't get adjusted to the move."

Brother John remembers how one day his aunt had drawn him to her, hugging him: "I know what's wrong with you, boy, you got to get your head together, John. Now, I'm your Aunt, and these men are your Uncles, even though you didn't know about them or never saw any of them before. Remember Mahalia Jackson who met you an' talked to you down in Greenville? Well, my friend Mahalia is comin' to church tonight with the Johnson Singers, an' I just know you'll want to see her again, to hear her sing.'

"And that night, at the Ebenezer Baptist, it was packed like a sardine can, about 2500 in the congregation. It was something! The Quartet was: Prince, who played the piano, and his brother Robert, who sang; Louise Lemon was the high soprano, and Mahalia, the contralto. Mahalia was accompanied by a jazz piano, which was a common practice in those days. People, that night, never before heard this new style of singing gospel. It got them all excited; it was the Sanctified way. And no one had ever heard a voice like hers in that congregation, 'cause the Quartet had come out of the Salem Baptist Church."

After the service and performance of the Johnson Singers, John accompanied his aunt and uncles onto the stage, where they greeted Mahalia. John stood shyly next to his Aunt Carrie and Mahalia recognized him, hugged him and said, "Well, look who's in Chicago. Am I glad to see you!" He grinned as she looked over to his Aunt Carrie, and added, "I know you're glad to have him with you." His aunt made a face in reply and turned away to talk with someone else. Mahalia got the message that all was not well, and urged John to visit her anytime after school. He promised that he would.

It wasn't long before John learned that his Aunt Carrie worked for Al Capone, the gangster who controlled all the prostitution and gambling rackets in Chicago. She was his maid and often served meals to his guests. John's remembrances of that time were more the emotional grab bag of a young boy's impressions of the bewildering happenings around him, as talked about in his Aunt Carrie's kitchen, rather than the actual history that went on in the Chicago white world. What the boy ingested was: "The Mayor of Chicago, John Cermak, announced that Capone had to get out of town. The Mayor said that someone else had to go too—a black man who lived on the South Side, who called himself 'Gloria Swanson,' like the movie star. He was a female impersonator. So Capone and 'Gloria Swanson' left Chicago together, I used to see 'her' marching around where we lived. Capone had some sort of thing going with her. I'd leave for school in the morning, and there'd she be, outside Carrie's house. There were two big white men guarding her. She looked like a big brown-skin colored woman, that's what she looked like. Later on, she got mixed up with Dutch Schultz, another gangster. Cermak was mad because Capone had put 'Gloria' up to making a dirty, nasty song about the Mayor. That's why he kicked them both out of town. Sometime later, Cermak was shot to death while driving with President Roosevelt in a parade in Miami."

That is what the boy remembered; there was no talk in Aunt Carrie's kitchen, evidently, of Elliot Ness or the Untouchables, a Chicago riddled with crime. Capone was Aunt Carrie's meal ticket, that was what was real. In the ghetto of the South Side, Aunt Carrie probably had celebrity status— she worked for the most powerful man in Chicago. But the boy also wasn't happy in Aunt Carrie's house. The fervor with which she had plucked him out of Greenville seemed to have disappeared. Living with Aunt Carrie was a disaster for both of them. She regretted bringing him north, it would seem. Perhaps it was his moody, artistic personality, his sensitivity. She was turning into his mother's twin—a cold, uncaring, and domineering woman. As for the other aunts and uncles he had been promised, his kin, they were equally indifferent to his welfare. For eight months, John was lonely and sad and missed Greenville, and Miss Bea's bordello, the only home he had ever known, a place where he had been an essential part of the scene. Miss Bea and Kersh had exploited him and didn't care if he went to school, but he had felt he *belonged*.

He turned toward Mahalia. It became his custom to drop in at her beauty salon after school, to see what errands he could do for her. She would ask him to shop for her while she was working, to buy pork chops and rice; she taught him how to prepare and cook an occasional meal for her and Ike, and soon the meal included him. If he had dinner with them, he would often stay overnight, even if the only place for him to sleep was between Mahalia and Ike in their double bed!

Aunt Carrie was aware of this change in his living habits, his being at Mahalia's a lot. She began to imply that she wouldn't mind if he left her house altogether; her husband, a church deacon, and their daughter, who was in high school, didn't seem to miss John when he was absent, either. Out of sight, out of mind. He was getting the message that he was not welcome anymore. Finally, it became no longer implied but asserted by Carrie that she wanted him out of the house. John was frightened, had no money, and knew of just one place he could go. He sought out Mahalia at the church before an evening service where he knew she would be singing. When he told her that Carrie had asked him to leave, Mahalia said her home was his home; he should collect his clothes and after the service was done, come to her and Ike.

Ike wasn't exactly pleased with the threesome in the same bed, but there was no other solution for the moment. He agreed with Mahalia that the boy had to have a place to stay, and Mahalia would have none of the boy sleeping on the floor. For John, being between them was heaven. Never before had he been so secure or felt so much concern. He now had a real family.

Mahalia's Aunt Hannah eventually located a folding cot for John, and there he slept for a few months, three feet away from the double bed. He insists he slept in a state of blissful insensibility, even if the new arrangement didn't add much to Ike and Mahalia's privacy. Ike was to prove himself a thoughtful, caring surrogate father. Whenever he had extra change in his pocket that he could spare, he gave it to John. He was curious about the boy's studies at school, and when Saturday came and Ike was off to the race track, he would take John with him and treat him to ice cream and other goodies. For the first time, John experienced what it meant to be a child and cared for. But the cramped quarters couldn't go on forever, that was obvious.

Though not more than a school boy, but having had adult chores and experience thrust on him from the time he was seven, it didn't seem unusual that he was offered a man's job by a white couple named MacIver, or that he would accept it, with Mahalia's approval. The MacIvers owned a small apartment building on the South Side and were looking for someone to take care of it and live in. Elizabeth MacIver took to John and offered him five dollars a week and his own place, a small bedroom, bath, and kitchen. A grown-up job, a place of his own, Mahalia and Ike nearby—John had fallen into a honey pot. The MacIvers were good to him. When he admired a set of cut-glass crystal objects she owned—a sugar bowl, salt-and-pepper set—Mrs. MacIver wrapped them up and presented them to him. "You have good taste, John," she said. "They're a gift from us. Real cut glass!"

His new acquisitions were such a point of pride that, when he heard that his mother was coming to Chicago to stay with relatives, he decided to present the glass to her as an offering—a kind of forgiveness for her having abandoned him. Their meeting took place at his Aunt Carrie's. When John presented his mother with the cut glass, and watched the pleasure on her face, he was thrown into total bewilderment by Aunt Carrie's reaction. She began to shout: "These things ain't yours, boy! Maggie gave them to me." She was saying that another aunt of John's, by marriage, had given her the glass. He couldn't believe his ears. "That isn't true," he protested. "How come Maggie gave you these? Mrs. MacIver gave me these as a gift. Maggie doesn't know nothin' about it," he recalls saying, and then his aunt's frenzy. "She came close to me and screamed: 'Oh, you sissy! You little dumb son-of-a-bitch! I'll get my gun and shoot your brains out!' I couldn't believe my ears."

John had no explanation for his aunt's insane behavior. He did expect that his mother would come to his defense, but all he could dredge up out of past remembrance was her response of: "Don't you two carry on like this, you hear!" He only felt shock at her unwillingness to defend him, since he saw in her eyes that she knew he was telling the truth, and her sister, for some reason, was lying.

"When Carrie's outburst occurred, a high deacon of her church, the Ebenezer Baptist, a Mister Walker happened to be in the apartment, wallpapering for Carrie. He was standing on a high ladder and turned to face Carrie with a shocked expression on his face, and pointed an accusing finger at her. 'Mrs.

Ferguson, how can you talk to your nephew like that? He's just a boy! And you're supposed to be a great woman at Ebenezer. How can you say such terrible things to him?' Carrie looked at him turned about on the ladder and spit out: 'This little bastard! I can't stand him any longer! That's why!' Mister Walker didn't say another word, but slowly came down the ladder, folded it, gathered up his paste pots, walked out the door and never returned."

Even though John was living and working at the Mac-Ivers', he still stopped by Mahalia's salon after school and often joined them for dinner. Mahalia taught him her cooking secrets, and the threesome continued as a sort of family, even though he wasn't living with them.

One afternoon Mahalia told him she wanted him to go with her to a church and join her in singing. That pleased him, because he knew he had a good voice and could earn extra money working alongside Mahalia, who was steadily rising in local popularity. John knew he had a future as a singer of gospel when the Reverend Bobby Williams of the Salem Baptist Church printed a handbill for one of the nights John sang alone without Mahalia. The handbill said: "Fire falls from Heaven when little Brother John sings!" Perhaps he could even make it alone, without Mahalia, if he had to. John was also watching and studying the blues style of Big Bill Broonzy, Chicago's star blues interpreter who hung out at a then famous saloon called "Big Bertha." Broonzy favored him and taught him songs and technique. Over the years the two men would became colleagues and friends.

In the early years, Mahalia objected to the singing of blues for herself and John, but later a time would come when she would be given the opportunity to perform in France for the first time, because of Broonzy's connections there, and would abandon her resistance to Big Bill and his music, though she never referred to her professional connection with him in Paris, when talking with friends about "the old days," even if she had traded on his Parisian reputation. (Years later John and Big Bill Broonzy performed together for a year in Paris, with great success.)

In John's early Chicago days, there were times when Mahalia, for one reason or another, decided not to keep a singing date. She began to feel comfortable about sending John, at concert time, with an excuse that Miss Jackson wasn't feeling well, and he would sing in her place, realizing that he was on his way up in his own right.

Looking back, Brother John, as he became known in church circles, summons up his past, when he traveled with and away from Mahalia Jackson, as a very capricious trip: "It was Mahalia, and for a time, Ike, who gave me affection I had never gotten from my parents or family in Chicago, but it's another absolute truth that in her later years of money and power, she became a devil of a woman. Money was to ruin her, and the whites who heard her perform ruined her, too. But that needs a little explaining. In the early days she was kind to me and to other people, she did good things for the church. When lightning struck, and she began to make money so fast, because of white acceptance, her head swam, she changed altogether. She became hard and domineering, impossible to live with.

"She moved away from Ike, though she saw him from time to time, and delayed divorcing him for many years. Along the way, I used to criticize her for saying or doing things I thought wrong. I dared to criticize her, because she had been my idol, she was so good to me when I was growing up. Where was that Mahalia? Lost, gone. All filled up with being the top, the best. 'You're bitter now,' I told her.

'You're cruel, devious, you won't sing for nothin' anymore, money, money, money! You set people up, stand back and watch them fall. Gifts and things come without repentance, Mahalia,' I reminded her. 'You don't have to be a minister or religious leader to lead people.' Those who have a gift, as she had, have the same responsibility to lead. And there she was, one time when we were going at it—on the thirty-third floor, double condominium she owned, and I said: 'You're up too high! Where is the Temple to God you said you were going to build with your money? Is this it?' She'd call me an ungrateful son-of-a-bitch, told me to mind my own business. I didn't care what she called me, we were very close and I knew I was right. She'd scream 'Bastard!' She hated to be reminded of her start-up days, the struggle and the sacrifices, when she wasn't playing so grand, when she was a different kind of person."

Many of the ministers, after auditioning Mahalia in her "start-up days" in Chicago, objected to her stage presence—there was too much body English for the taste of their congregations. She had a tendency to shake and twist about—movements definitely inappropriate for those in the pulpit. Her sensuous power disconcerted the generally conservative congregations, their very sober ministers and elders. Unless she agreed to wear suitable robes that concealed her vaulting feelings, they said, she would risk forfeiting church engagements, maybe even jeopardize a career, if it came to singing in their churches.

When she did put on the robes suggested, she ascended to the performing pinnacle of the gospel circuit like lightning. Invitations and large fees were now thrust upon her. Black audiences way beyond Chicago's were becoming quickly seduced by her injunction "to make a joyful noise unto the

Lord." In her deeply individualistic manner—running and skipping down the church and concert hall aisles, her eyes closed, hands tightly clasped, with feet tapping and body throbbing, all the while her voice soaring as if there were no walls to confine its spiritual journey—she was utterly possessed and possessing. It was pure theater.

"Move On Up
a Little
Higher"

Louis "Studs" Terkel, Chicago's long-reigning, celebrated raconteur—disc jockey, author, sometime actor—had started his radio program called "The Wax Museum" right after V-J Day, on station WENR. From the start he had a mission. Not a reticent man, Studs salt-and-peppered his musical programs with wit and trenchant social scene commentary—a persuasive good neighbor who, even if you didn't agree with him, you didn't turn off. From the very beginning there was a buoyancy, a freshness about Terkel's commentaries; politicians could come and go, but he was to go on forever—to this day, more than forty years later.

In the beginning, his audiences were mostly white, but blacks were soon drawn into his unorthodox web of broadcasting, for it was Terkel who was the first to introduce black entertainers and ethnic music on the white airwaves in Chicago.

The way he tells it, he was dawdling in a record store on Michigan Boulevard one lunch hour, when he heard a woman's voice singing a gospel, probably (he thinks), "I'm Goin' to Tell God All About It One of These Days." Mahalia Jackson was unknown to him, but the power and majesty of the voice knocked him out and still does, he says. "I've been movin' on up with her ever since." He wanted to meet this woman who lived somewhere in Chicago and did a bit of detective work with the Baptist and Sanctified church ministers. They quickly identified the voice for him.

Traveling around the black church circuit of Chicago, he lost himself in the music and stirred-up congregations, watched Mahalia perform, and offered her an opportunity to appear on his program—she could sing whatever she liked. Studs paved her way with his audience, by announcing his discovery: "There's a woman, my friends, I've seen and heard, who sings like the great blues singer, Bessie Smith, only Mahalia Jackson of the South Side doesn't sing the blues. She sings what is known in her church as 'the gospel.' All right, friends, now listen to this!" And Studs played the 78 record of the gospel he had heard in the music store—"I'm Goin' to Tell God All About It"—played it again and again, until the grooves wore thin. It was the only recording she had made at the time. He promised his audience that as soon as he could make arrangements he would bring Miss Jackson to the mike, live. It was those first broadcasts of her work, her chatty talk with Studs Terkel that introduced her to the world beyond the black church community.

Terkel paints the style of Mahalia's gospel singing in a breathless language of admiration and affection: "Watching her in a church, particularly a church where poor people came . . . her relationship to the congregation was something to experience. You didn't forget—the call and response, the give and take; she didn't sing with her voice alone, it's the body, the hands, the feet. And she used the phrase 'demonstrating' which has several connotations. 'Demonstrating' is reflective of life itself. She explained to me that plain spirituals are slavery songs. Gospel is post–Civil War. Mahalia would sing both, but gospel had the bounce though there were spirituals with life in them too, as well as the sober drive, the singing drive toward freedom. She explained to me that the spiritual wasn't simply about Heaven over there, 'A City Called Heaven.' No, the city is here, on Earth. And so, as we know, slave songs were code songs. It was not a question of getting to Heaven, but rather to the free state of Canada or a safe city in the North—liberation here on Earth!"

Studs Terkel's timing was stunningly correct about what kind of sound was needed in the post–World War II years— the up feeling, an affirmative cry for peace and prosperity. With the gospel sound, all would be, could be, right in the world, and God was in His Heaven. There it was, a new beat, a musical stirring that white ears had never before heard—the celebratory life force, the segregated healing sound that had sustained black people in their anguish and separation from the mainstream of American life. Jazz and the blues had been embraced by whites; now it was the time for gospel, and the sound that Mahalia could create so magnificently.

Bess Berman, an adventurous woman in the recording business in New York, fell in love with Mahalia's gospel voice and signed her to a recording contract in 1946. Apollo Records was small but successful in introducing talented musicians, many of them black, to the black market, names like Dinah Washington, Charlie Barnett, Billy Daniel, Woody Herman. Berman made money from blues discs and later expanded into a rainbow of folk and popular music—Calypso, Western, Latin, Jewish, Gypsy, and Josh White, with his black work songs, "a bit more urbanized," the raw pain of chain gang and cottonfield smoothed out.

Bess gave Mahalia a guarantee of $10,000 a year, and Mahalia promptly recorded four sides, including: "I'm Goin' to Tell God All About It" and "Wait Till My Change Comes." When their first sales report was released, the sales figures were so meager that Mahalia fell into a depression for weeks. She said she wanted *out*, she told Bess; this recording stuff was a flop. Gospel, she was certain, wasn't commercial because white recording companies couldn't find a market with whites. Anyway, how could whites understand black church music? Bess thought about Mahalia's reaction and the poor sales report. She told Mahalia she didn't want to put out any more money for gospel if there wasn't a market for it. Mahalia almost conceded defeat, but she had one final song she wanted to record, if Bess would go along. They talked it over on the telephone many times, then in person over a meal whenever Mahalia came to New York for recording sessions or personal appearances.

Back in the thirties, a Reverend Brewster in Memphis wrote a gospel he called "Move On Up a Little Higher." A version was performed by Queen C. Anderson at a National Baptist Convention, and an organist in New Orleans, Professor James "Blind" Francis, gave the song many enthusiastic performances and recorded it. Mahalia was certain she could add her unique swinging style to the song that had not moved far out of black church circles.

Mahalia performed the song for Bess, a last-ditch effort to chose a gospel that would succeed in breaking through the public apathy to her commercial offering. The song was so long in its structure that Bess, now a zealous partisan of Mahalia's, agreed to produce it on both sides of the then 78-rpm record, as parts I and II. It was still a gamble, but Bess, by now, was confident the gamble was worth taking.

One of these mornings, one of these evenings . . .
　　I'm goin' to lay down my cross and get my crown.
Late in the evening, I'm goin' home to live with God on high.
As soon as my feet strike Zion, I'm goin' to lay down
　　My heavy burden . . .

Put on my robe in glory and tell my story . . .
　　How I come over hills and mountains . . .
I'm goin' to drink from the crystal fountains . . .
　　And move on up a little higher!
All God's sons and daughters will be drinkin'
　　That old healin' water . . .
And I'm goin' to live on forever
　　And meet old man Daniel . . .
Put on my robes in glory
　　And move on up a little higher!

The gospel, indeed, proved to be "roaring trade" for both the artist and her recording company. And every time Mahalia sang her "trademark," she'd invariably improvise the lyrics and creatively fracture the gospel's structure.

The results were spectacular, the recording an immense success, and it produced royalties of over $300,000 for Mahalia in the first year. In the face of Bess Berman's success, her husband, a jukebox operator in Miami with Mafia connections, contracted with his wife to release "Move On Up" for his jukebox empire. The results were, again, extraordinary. He had Mahalia come to Miami to sing before the Juke-Box Operators Convention. Everyone knew her song so it was a warm celebration for her.

What Mahalia didn't know was that Rosetta Tharp, a competitor in the gospel field, had a contract with Jukebox Berman that preceded Mahalia's, and Tharp's contract included veto power on any gospel competitor they might want to sign. Mahalia's "Move On Up" had already been manufactured into the system, but after that she was summarily cut out. Mahalia was outraged with Berman's bowing to Rosetta Tharp's command, but she recorded no other jukebox numbers, though the bar-and-tavern crowd kept asking for more.

With "Move On Up," like Times Square's impossible-to-ignore electric traveling sign, flashing news, Mahalia was a rising personality, attracting the attention of freelance agents and the William Morris Agency. She decided, rather, to stay with black impresarios on their way up, dealing with church-only personal appearances. At least she could trust them, and they, being black, could comprehend her distrust of *out there*, her unorthodox money requirements: I sing? You pay me in cash the same day.

Black clubs in New York and Chicago were now paying her $1000 and more a night, payable in cash by the close of the performance.

The Golden Gate in Harlem was redesigned, from a saloon to an auditorium with a couple of thousand seats, to accommodate the galloping gospel aficionados. It became the mecca for gospel and jazz, starring Duke Ellington, Count Basie, Rosetta Tharp, and her competitor, Mahalia Jackson. The swinging citadel bought radio time and put a crowd of sandwich-board men out on Broadway, to stir interest in the midtown crowd to travel up to Harlem where it all *was at.*

By 1951, Joe Bostic, Mahalia's church agent, was expanding his horizons for his gospel clients. He wanted to take the biggest gamble of all and rent Carnegie Hall. A Saturday night with his talents would rock it! He hocked everything he owned that was movable to pay the rental fees and program costs and gambled that the night would be a sellout—and it was. Part of the success of the evening must be laid at Mahalia's feet; she talked herself hoarse with reviewers up front, spent her own money wining and dining people, beating the drums for the evening to come.

As she put it, inelegantly: "I worked my ass off. I always do. No one could ever get the attention I got, because I used my own money to get attention. And that's that!"

She had shared the stage with Rosetta Tharp, and the Gaye Sisters, Clara Ward, and some lesser-known lights. But it was Mahalia who was luminescent.

The *New York Times* man of the day admitted that he thought he had confronted a Cecil B. De Mille mob scene when he reached Seventh Avenue and 57th Street. And inside Carnegie Hall, an excited crowd wouldn't sit in their seats. They roved the Hall, swarming up and down the aisles. Everything was sold out, from the boxes to the last row in the second balcony crow's nest. The Golden Gate had moved to Carnegie Hall for one night, to mix with Manhattan's new audience for the gospel, and the coming together was astonishing.

Mahalia remembered the evening: "When I came out onto the stage to begin my program, there just wasn't enough room for me to stand next to the piano. I looked over at Mildred, my accompanist, and shrugged. We'd just have to do as well as we could, with people crowding our space. When folks, during my numbers, got to their feet, danced in the aisles and the ushers were helpless to stop them, I tried to say something, and folks cried and stamped their feet. Even I got carried away. I found myself down on my knees, yes, singing my heart out. Then, when I got to my feet, I told the crowd that had come all the way from Boston to Baltimore, that *we* were in Carnegie Hall, and if we didn't cool it, the authorities and the police would put us into the street. I guess they heard me 'cause most of them settled down and the program went on to near midnight."

That evening broke all house records, even those set by Toscanini and Benny Goodman. The reviews were raves, and Mahalia got the overwhelming kudos. She was walking on air, but confided to close friends that before the performance, she was filled with apprehension. "I was scared to death. I was up praying all night long . . . and prayed so hard to get back my courage and self-consciousness." (What an apt misuse of the word. She undoubtedly meant, given the context of that evening and the crucial test she faced, consciousness of "self.")

"I'd heard," she added, "and known that Carnegie Hall was the greatest place to sing in America, because of who had appeared there. Like Caruso, Robeson, Anderson, and Lily Pons!

"I liked being up there with those folks, so I got over my fears and went on, and it was all right!"

The Image
Respected

After Mahalia and I had met in New Orleans in 1955, we went our separate ways—she to Chicago, and I back to New York, to search for production investors. I spent months on it, but failed to raise the money. Mahalia Jackson, the little girl from "back' a town" New Orleans, had already flashed across the sky, a hellbent-for-heaven gospel-singing meteor; had already stormed Carnegie Hall; and had moved over from Bess Berman's Apollo Records to Columbia Records. But the people I approached for financing were unconvinced by my arguments that Miss Jackson, now being touted by Columbia Records as "the world's greatest gospel singer," would attract a film audience. I was being accused of "having a very personal fancy." The trouble was that none of them had heard gospel music, understood how it differed from jazz or blues or ragtime, or had any idea how important gospel was becoming, with a cash register sound. Profits aside, whites were ignoring the sparks of a civil rights revolution about to erupt.

Mahalia had a voice and presence that presaged the "Black is beautiful" gauntlet, a voice requiring attention and redress of ancient grievances, but whites were deaf to this new energy. The unanimous opinion of my film contacts was: Any film starring that black lady was certain to flop, so why invest? In their view there was no market out there for it, except a black one. I had to give up my dream of making "Kinfolk," the three-part feature of American roots—it was an impractical dream. I turned my attention to television assignments.

Mahalia, I know, was disappointed in my failure to raise the money; she told me so, bluntly and aggrieved for the moment. She might have been disappointed, but I was heartbroken—a state of mind film-makers often have to face in their sometimes artless business. Mahalia didn't shed tears for long; she had bigger fish to fry, for her career was zooming.

Though I hadn't come through for her, curiously, she trusted me and my instincts. We had spent many days together in both Chicago and New Orleans, reliving her personal history of blackness, and I had survived her private test. Unschooled and lacking alleged sophistication according to the standards of white society, she had fierce instincts, dead-right, and exquisite insight about everything that had to do with her career, talent, and the circumstance of being a black woman.

She kept in touch and began to ask me to oversee some of her television programs originating in New York, particularly when it concerned how she was lit and costumed. In television of the 1950s, directors and camera technicians hadn't learned or wanted to acknowledge that black skin had to be illuminated differently and with subtle shadings, or the planes of the face would be lost. Whether it was racially motivated or just plain technical insensitivity, the sterotype of *darky*, the expected Negro image of the "mammy" or "Aunt Jemima," was supported, with the rocking chair on a rural porch, the scarf or bandana around the head. The image was never questioned; it just happened.

I would precede Mahalia (and Mildred Falls always at the piano or organ) on a shooting day at a studio, and warn unsuspecting producers and their staff that Mahalia Jackson would "walk" unless they modified their traditional lighting scheme and production style, unless they didn't mind being guilty of what appeared to be dogged racism. Staff and producers tended to resist such advice, for, who the hell was I and who the hell was Miss Jackson to affect "the standards and practices" of the industry? But our demands did force a rethinking of craft and racial sensibility, at least as far as her appearances were concerned.

On one of those New York television trips of hers, in the dog days of August, Mahalia called me to discuss a series of appearances she would be making. It was an unbearably hot and humid night, the kind that sends everyone who can fleeing to the country or shore, and those who can't thank God for the gift of air conditioning.

Mahalia asked if I was free, and whether I would join her and Mildred Falls for dinner. I suggested I meet them at the midtown hotel where they were staying; we could eat in a nearby restaurant. The women, both large and heavy, were awaiting me, dressed in immense, white shifts—intimidating, outsized angels. Ensconced in a large bed-sitting room, Mahalia preferred to remain in the air-conditioned suite and have dinner sent up. Knowing something of her eating habits, I suspected she wouldn't much care for the hotel menu, but I phoned down to the *maitre d'* for the menu to be sent up. A young waiter arrived with it, and Mahalia took it from his hand and went to confer with Mildred, sitting on a bed, while the astonished waiter looked on. Here was a single white man in the company of two very large black women dressed in what appeared to be nightgowns. We must have presented an unusual tableau, suggesting an unusual divertissement.

As I suspected, Mahalia was annoyed that the menu didn't include ham hocks, greens, black beans, and rice. I tried feebly to justify the standard midtown, middle-class white fare, but got nowhere. She resented that the menu made no provision for "people of color." "I bet you want somethin' like a fruit salad or a tuna fish spread, white folk's food, bet ya'," she ridiculed, and I agreed. The heat and humidity called for nothing else; the fare didn't have anything to do with being black or white, but instead with life or death. Now she decided to amuse the waiter. Had I ever eaten stuffed raccoon or baby alligator? I hastened to say I had made a Bell Telephone Hour television show in western Louisiana, in Cajun country, and, as a guest of those folks, had tried some marvelous Creole food, gumbos of okra and rice, crayfish étouffé, red fish, even wilder foods caught in bayou traps, I boasted, but that wild things were not my favorite food.

The women eventually made their choices: corned beef and cabbage, with a side order of French fries. And I ordered a cold shrimp salad. We conferred about my role in helping make arrangements for the TV show she was scheduled to do in the next few days. I was at her service, Mahalia knew that, and without payment of any kind. It was my way of compensating for having disappointed her—my failure to do the film,

I suppose. But it was also a point of honor—an opportunity for me to confront the establishment, and Mahalia went along with me, tacit and willing, particularly about having her own producer without payment.

With dinner over, I called downstairs for the table and plates to be removed. In no time at all, there was a knock at the door, as if our young waiter had been waiting outside. There he stood, surrounded by many busboys. They hastily removed the dinner paraphernalia, all the while giving our trio a close inspection. Our young waiter had obviously shared his curiosity about the odd mix in Room 1536. Even in that hotel where everything and anything went, it was something not to be missed.

Turning Point

One hot summer weekend in 1951, Mahalia and Mildred were driven up to Lenox, Massachusetts, in the Berkshires, as guests of the Music Inn, a prestigious summer school of the Institute of Jazz Studies in Greenwich Village. The Institute was the creation of the celebrated musicologist Marshall Stearns, whose seminal book *The Story of Jazz* not only chronicled the West African drum origins of jazz music, but was probably the first exposé of the racist history of the 1930s black and white jazz bands. Stearns put it this way: "If Benny Goodman became the 'King of Swing' in 1935, reaping all the publicity and profits, the man behind the throne was Count

Basie. For it was the Basie band that gave depth and momentum to the whole swing era while planting the seeds that later gave birth to bop and the 'cool' school of Jazz.'' White bands of those years, Stearns wrote, were applauded for simulating black band performances and music, while black musicians generally were excluded from playing in such groups.

While Music Inn summer session housed a celebrated faculty of professional jazz performers and savants from the Juilliard School of Music and other prestigious music schools, it lacked any artist of the gospel. When Stearns became familiar with Mahalia's Apollo records, he knew he was about to close the gap by inviting Mahalia Jackson and her accompanist, Mildred Falls, to spend a weekend at the jazz "think tank."

Years afterward, recalling the week that grew out of that weekend, Mahalia gave a belly laugh: "'Well,' I said to myself when Mildred and I got to the Inn—'Mahalia, you finally made it into the white folk's world! Look where it's landed me and Mildred!'" She was referring to the rustic-styled room assigned to them by the Music Inn staff. It hadn't taken her long to discover that the room had once been a part of the stables of what was originally a horse farm.

Mahalia and Mildred provided a gospel feast for Stearns and his guests, with some rousing bonfires of their favorites, like "Move On Up" and "Didn't It Rain." For four hours Mahalia adroitly fielded the searching questions of the professionals in the audience, as well as venting her distress over the growing commercialization of gospel.

"Some of the market out there is falling into the hands of hustlers and sharks in the business," she lectured. "You know there's plenty of weak-headed gospel singers around, and many of them can be charmed off their feet into helping make gospel singing 'entertaining.' Now I hear a lot of sound, like it's half jazz. Well, gospel isn't! The way I see it, if you sing the gospel, you never need any artificial *anything*. . . .

"Now you know there's nothing wrong in being commercial, but the work has to inspire. The way I see it, I don't care if a man is a gambler or a thief, or even a murderer—every man and every woman has got to believe in something, something they can look up to. So, I believe that the gospel, and the singing of it, can be both commercial and uplifting at the same time." She was saying in her own inimitable way: there's nothing wrong with selling gospel in the marketplace as long as you're selling it pure and untainted.

After dispensing with the business side of the gospel, Mahalia rapped anecdotally about the spiritual and racial aspects of the art, how black gospel had evolved from white gospel in the post-Emancipation years of the rural South, influenced by the rhythms and call and response in church services, combined with the improvisations of African music. She spoke of the origin of southern tent shows that had provided entertainment for rural audiences, black and white, when there were few other distractions for isolated and poor people. She recalled that for many years, beginning in the mid-1930s, vocal ministers and board members of black churches opposed, resented, were even horrified by the descent of the gospel into songs of broad popularity. It wasn't until Mahalia and other gospel stars had found white allies that concessions were to come from black churchmen. They now had to admit that a popular market existed for commercial performers and composers of the art. Then Mahalia warned her benign audience in the Berkshires that Jim Crow had to go, or there would be a political calamity in the land. With pride, she ended her lectures on the gospel, acknowledging that she was a self-taught singer: "I just found myself doing what came natural to me."

Later, Stearns made notes in his diary concerning her contribution to the week: "She breaks every rule of concert singing, taking breaths in the middle of a word, and sometimes garbling the words together. But the full-throated feeling and expression are seraphic!"

Years later she thought of the effect she had made upon her famous audience at the Music Inn: "Wow, I just about drove all those folks crazy when I got around to doin' 'Didn't It Rain.'" In her memory, that weekend that had evolved into a week was to be one of the consummate periods of her life.

Following her lecture, a tall, lanky man, with a toothy radiant grin and sporting a crewcut that was his trademark, pushed through the crowd that surrounded Mahalia and Mildred. He took her totally by surprise, leaned over and kissed her on her cheek. "You were simply wonderful, Mahalia. Just superb!"

The man was the most celebrated and assertive executive in the record business of the day, none other than John Hammond—passionate about jazz and blues, gospel and folk; critic and recordings producer for Columbia, Vanguard, Mercury, and numerous smaller labels; producer of 78-rpm singles and LPs between the early thirties and the late seventies. He was a backer of Café Society, the renowned nightclub in Greenwich Village owned by Barney Josephson; and along with Max Gordon's Village Vanguard, the two nightclubs were first in the U.S. in the breaking of the color barrier.

For thirty years and more, Hammond asserted his talents and taste by serving as an impresario and recording producer for such immortal luminaries as Billie Holiday, Count Basie, Teddy Wilson, Lionel Hampton, Aretha Franklin, Bob Dylan, Bruce Springsteen, and briefly, as it turned out, for Mahalia Jackson and Brother John Sellers.

In the mid-1930s, Hammond's sister Alice married

Benny Goodman, a promising rising star. Though some evidence suggests that they had been lifelong friendly enemies, Hammond served Goodman variously as critic, talent scout, and tour manager for the rest of their lives.

Hammond had the resources to live and work in relative ease. He often accepted employment in the music business at relatively small salaries, while counting on his contractual percentages to bring him profits as the result of an unerring taste, if not luck, in choosing "winners." Once, when a nosy fellow approached him at a cocktail party to ask him how he could afford his life-style, Hammond gamely replied: "The Hammonds never had money. They married it!" On his mother's side, his grandparents were the Vanderbilts who controlled the New York Central Railroad, some public utilities, and W. & J. Sloane Co., the elegant home furnishings store on Fifth Avenue.

John Hammond got along famously in the recording business, enjoying himself the whole distance, even though he collected quite a few disclaimers along the way, either through natural competition or because he may have been a superior self-promoter in a business notorious for rivalry and fratricide.

There is little doubt, however, that he was an invigorating investigator who uncovered more talent in isolated bayous, river bottoms, and forlorn plantations than anyone else in the business; and for most of the artists he discovered he provided the ultimate in public exposure. His determined, successful promotion of black talent—to record companies and the public—earned him the tag of "The Great White Father," bestowed on him by his many black admirers.

Following dinner one evening at the Music Inn, Hammond sought out Mahalia, with Mildred always in close step behind her. "I'd like to talk to you about your recording future," he said. "Columbia is aware that as Bess Berman is in some financial stress, she may want to sell off her catalogue, and that includes your contract with her. We understand she guarantees you ten thousand a year. Columbia can certainly top that. Say, twenty-five, anyway. So, my dear, perhaps it's time for you and your advisors to think about your coming over to us . . . one of these days."

Hammond guessed that his colleague at Columbia, Mitch Miller, would offer her a five-year renewable contract covering four recording sessions a year. When she delicately inquired again about money, Hammond assured her that his estimate might actually be lower than the deal Miller would ultimately concede, to get her signature on a contract.

Mahalia's native acumen when it came to figures made her decide that her strategy with Columbia Record should be: Let the company, in the guise of John Hammond, chase after her. She told her friends and agents that if Columbia wanted her so badly, well, she'd just raise the ante for her services; black gospel singers shouldn't be expected to work at bargain rates! She was determined to sit out the time and listen carefully to the best offer that would come from this charming super-salesman, who, she admitted later, was very different from the many white recording promoters she had crossed swords with. Hammond was respectful of black sensibilities, and she believed that he was utterly free of race prejudice. On that count alone, he was way ahead of the other petitioners she characterized as "hypocrites."

The wooing of Mahalia Jackson by John Hammond had begun. "I promise," he told her, "Columbia will promote the hell out of you." He titillated, and she enjoyed it. "I'm going back to work for the company soon, and you can be sure I'll oversee the deal. You'll make more money, Mahalia, than you ever thought possible in this world." He assured her many times that the deal would guarantee that she would be the only gospel artist on the Columbia roster, that the publicity in the trade papers and on her record covers would identify her as "The World's Greatest Gospel Singer!" And so, indeed, it came to pass, though it would be a few years before Hammond returned to Columbia Records.

By early 1952, the *entente cordiale* between Mahalia and Bess Berman had shattered. To justify breaking her contract with Apollo, Mahalia countered with not-so-coy accusations that she had been short-changed in her royalty payments. Bess screamed foul. Someone in Mahalia's coterie then suggested she get a smart accountant to file a demand with Apollo that he examine the Jackson account. Bess, naturally, balked at this threat of invasion and refused to cooperate. But Mahalia, it appears, had no stomach or cash with which to hire lawyers to bring Apollo into court. Instead, she called Mitch Miller, the contract negotiator at Columbia, to tell him: "I'm ready to consider your best offer."

Some months following the meeting of John Hammond and Mahalia in the Berkshires, he became a producer for Vanguard Records, to oversee an eclectic album entitled "Spirituals to Swing," based on a 1939 Carnegie Hall concert. It would be a couple of years before he rejoined Columbia Records, but he didn't forget his advice to Mahalia, to which he had a few more ideas to add. He telephoned her at her Chicago home, with an admonition:

"Mahalia, I'd like to give you some second thoughts, in case you're getting close to a decision about Columbia. Let me remind you, my good A&R friends at Columbia really know next to nothing about gospel. In fact, I don't honestly believe any of them even care enough to listen to your Apollo recordings. And so, I've been thinking that you could lose the best audience you've ever had: your own black audience, church-grounded. While you're reaching out to find that elusive white crowd Columbia expects will come to swoon over gospel as it came to revere jazz and the blues, *it just might not happen*. So, my advice to you, my dear—think about this before you leap."

She thanked her new friend with the crewcut for his advice, though she had no intention of taking it. Her desire to penetrate the commercial stratosphere was too strong, covetous; she was on a roll; she had no time to waste; she knew she had what people wanted. She would listen to no advice but her own.

But, curiously, when Mitch Miller sent her a contract, she didn't sign it. Plagued by anxiety about making the right turn in her career, she consulted Robert Ming, her trusted Chicago Loop lawyer. After reading the terms of the Columbia contract, Ming recommended that she sign the papers and return them to Miller, assuring her that it was a good deal the first time 'round, and in four years the renewal would undoubtedly be more attractive.

But still she put off signing the papers.

During the winter and spring of 1953, Mahalia and her entourage took off on a performance tour through Memphis, Little Rock, Dallas, and Oakland, California. She carried the Columbia contract with her, stuffed into her big purse. At every stop-over, in between concerts and church appearances, she obsessively pored over its contents, some of the contract provisions being confusing, but she was determined to understand the language and stayed up late into the night re-reading the fine print by the light of low wattage, hotel bedside lamps, while her friends slept in adjoining rooms.

It was Easter Week when they were in Oakland, giving concerts and making personal appearances. Near dawn after

Easter Sunday night, sleepless, exhausted, Mahalia looked up from her contract papers and sensed, as she put it, that her Lord had materialized in the hotel room and gave her a sign, finally, to put her signature on the Columbia contract. Shaking with excitement, she tiptoed from room to room in search of a pen, not wanting to waken anybody. When she found one, she returned to her room and scrawled her name on the dotted line, then fell back on her pillow into a deep sleep. Monday morning, she sent some Oakland friends off to the post office to mail the contract back to Mitch Miller.

Two and a half months later, Mitch Miller countersigned the papers. He then telephoned a friend in Chicago, the producer Lou Cowan, agent-impresario, and asked him to create a half-hour radio show originating from Chicago, starring Mahalia Jackson singing gospel and spirituals. It took Cowan little more than a week to design the program and obtain management approval all the way up to the CBS chairman, Bill Paley.

When Mahalia learned that her CBS radio show would need a staff writer, she asked for her friend and champion, Studs Terkel.

The year was 1954. Studs Terkel was unacceptable to CBS, because his name was included in the infamous blacklist of the day known as *Red Channels*, a reactionary magazine subscribed to by studios, networks, and advertising agencies. Month by month, *Red Channels* listed the names of thousands of men and women who were charged with being disloyal Americans (read: "communists"). These charges were rarely, if ever, substantiated publicly, but simply by listing their names, people became unemployable in their professions of cinema, radio, TV, advertising, education, and government.

In Terkel's case, he had always been an outspoken critic

of Jim Crow. Though Chicago, the city he celebrated every day on radio, with music and social commentary, considered him to be among its first citizens, a bold slugger for justice, that, according to *Red Channels,* made him dangerous.

To her credit, Mahalia insisted that Terkel be the only candidate for staff writer on *her* show. CBS agreed to hire him, but with a single condition: His name would not appear in the program credits or in advertising. Terkel agreed. The impertinence of the network and the idiocy of the political climate of the day affected him like a fly on an elephant's back.

The half-hour "Mahalia Jackson Show" ran a total of twenty weeks. Seventeen of the programs were half an hour in length; the last three were ten minutes each. The show began September 26, 1954, and ended on February 6, 1955.

Mahalia sang traditional gospels and spirituals, and a sprinkling of miscellaneous songs that included the Brahms "Lullaby," "Silent Night," and "The House I Live In." Large, loyal audiences tuned in, and the reviews were luminous. But a major ingredient for the show's success was mysteriously absent—namely, a sponsor. Mahalia was too emboldened by the positive reviews and audience reactions to give much thought to the network's commercial concerns.

Journalistic rumormongers had been busy from the start, implying that the program would be in trouble once it sought a white audience, and there were others who suggested that Miss Jackson had very little more than a regional following and so great expectations were unjustified. If these not-so-quiet mumblings affected Mahalia, she never gave any hint of having heard them, much less being offended by the charges. When judgment day arrived, however, and the network officials slashed her last three programs to ten minutes each, Mahalia understood the message. Her show was going out of business! The condensed program from thirty to ten minutes must have cost her dearly in pride, having to compress her art and rein in her incredible energies for a humiliating ten-minute segment.

In New York, John Hammond was working for Vanguard and otherwise employed in the music business, but when he heard that Columbia had signed Mahalia and that she had begun to record at the old CBS studio on East 30th Street, he was so pleased with his matchmaking role he sat down at his typewriter and wrote a release of the event for the *New York Times:* "Another new development being preserved on records is the music of the Negro gospel singers, up to now available on obscure labels catering to a small audience. Within two weeks, the voice of Mahalia Jackson, one of the great religious singers of our time, will be available on LP."

About the same time as Hammond's news release, TV station WBBM-TV in Chicago invited Mahalia to make numerous guest appearances on the program "In Town Tonight." Until then, neither she nor any other black entertainer had been sought after by the networks, and, of course, it pleased her. But it would be a few years before the civil rights movement would begin to affect segregated studio programming, despite the fact that journalists such as Chicago's Irving Kupcinet were wondering out loud in print why such an original artist as Mahalia Jackson was not being invited to appear in the launching years of television—a gentle jibe at the industry's obstinate racism.

As for Mahalia, she again made a proviso that her friend Terkel must be employed as writer of the TV show when she appeared. Again, the station conceded, but again with the condition that the blacklisted writer remain anonymous.

A photograph from one of Mahalia Jackson's LP covers. Photographer unknown.

(Courtesy of CBS Records)

In the sound studio with Rosemary Clooney. (Courtesy of CBS Records)

Mahalia and Mildred Falls.

Goddard Lieberson, President of Columbia Records—from the "Playback Series," 1961. (Courtesy of CBS Records)

"Playback Series." (*Courtesy of CBS Records*)

"Playback Series." *(Courtesy of CBS Records)*

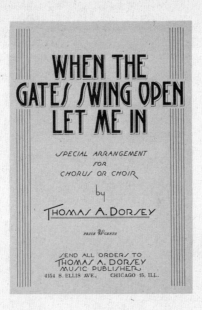

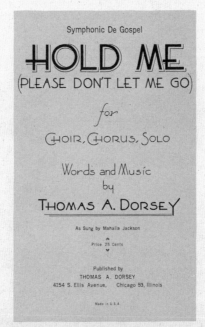

Two of the original scores by Thomas A. Dorsey favored by Mahalia.

Mahalia performed this piece in honor of her close friend Martin Luther King, Jr.

Mahalia with Hubert Humphrey. (Courtesy of CBS Records)

Mahalia Jackson was an admirer of Liberace, evident by the way she holds the musician's portrait; George Avakian is at left.

Mahalia with Harry Belafonte and Leonard Bernstein. (Courtesy of CBS Records)

Mahalia flanked by two unidentified men; George Avakian at far right. Photographer unknown.

Clockwise, left to right: Mildred Falls at the piano, Mahalia Jackson, an unidentified man, John Mehegan, Professor R. A. Waterman (Northwestern), Professor M. W. Stearns (Hunter College), John Hammond from Columbia Artists, and Professor W. L. James (Spelman College). (Clemens Kalischer, Institute of Jazz Studies/Berkshire Music Shed)

(Courtesy of CBS Records)

Brother John Sellers—from the author's personal collection.

(Courtesy of CBS Records)

Customary in black churches was the distribution of "celebrity fans" with reverse-side advertisements. This one memorializing Mahalia was provided by a local undertaker.

Mahalia moved into the TV and recording studios with great expectations. But she found it onerous, impossible from the very start. Management bombarded her with caveats, and she must listen to their advice on the selections of songs, orchestrations, and arrangements. She was not used to outside creative control in what she chose to sing or how she sang it. Never had she had to justify her songs or how she sang them to church ministries or the management of secular concerts. From them, she had received only love and esteem. Television and recording relationships apparently were different.

She talked about one such unexpected circumstance: "When I first walked into that CBS studio, I admit I was pretty much a total unknown. None of those folks had ever been inside a black church, or listened to gospel on early records. But they had their orders to work with me. It was *my* show certain nights, and they did work with me. But some of them may not have liked what I do!

"Well, that first day, the arranger walked over to me even before I had my coat off. He sat down on the piano bench and told me he just couldn't follow the song I'd chosen to sing for an opener. Now the song he was talking about is a very old and famous spiritual, 'Nobody Knows the Trouble I've Seen.' Of all the music in that studio, he should have known that song! Well, that man told me that the way I was going wasn't the way the song was supposed to go! Can you believe that?

"So I just talked right back at him. I said: 'How come, mister, you think you can tell me about that old song, when it was born in my mouth?'

"And, you know, he answered me back with ridicule. He talked to me real bad. But I held my tongue and said nothing more. I decided to answer him through the song itself, and the way *I* sing it."

Another time, a woman journalist stopped Mahalia on her way out of the studio. The woman asked her to explain the extraordinary success of the program. Mahalia thought it a silly question, but tried to give her a decent reply.

"How do I know why it gets a good audience? Maybe, it's CBS just now getting around to the gospel. And maybe, it's just news to a lot of the folks here up North. You see, everybody in the South has heard and knows the gospel, all their lives. I guess maybe it's just that simple, lady."

George Avakian was assigned to be Mahalia Jackson's artist and repertoire man at Columbia. They liked each other through the years, and it's not surprising considering the story he tells of seeing her perform for the first time in a San Francisco church in 1944 while he was a young draftee awaiting word of when he would be shipped out to New Guinea.

Avakian has the ebullience of youth as he recalls that fabulous night of discovery. He remembers trembling with excitement, breaking out in a sweat while watching Mahalia Jackson shake and rattle the old building with her spectacular vibrato. Even now, he believes it a miracle that after hearing her in a San Francisco church, in the middle of a war, ten years later he was to become her recording producer. "So much for a random adventure," he comments, "in which I found gospel music and its inspired oracle."

As for Mahalia's occasional irascibility, her temper when she was thwarted, Avakian recalls a somewhat cooler atmosphere than she describes. He insists that she alone chose her repertoire, that their working relationship was friendly and creative.

Though he represented management, Avakian was, and remains, a passionate admirer of her genius. He admits there were times—moments perhaps—when she gave evidence of feeling lonely and defensive in a stressful work session. But as for racial frictions between Mahalia and the staff, Avakian insists he was unaware of any. With the passage of thirty-plus years, and considering the delicate subject of racism, what white person, however personally innocent of bias, could imagine the blows, subtle or otherwise, sustained by a black artist in those days before the Negro became the proud and vocal black, before the civil rights movement picked up where the Civil War left off?

"I know for certain," Avakian maintains, "that Mahalia worked compatibly with Mitch Miller, and my producing associate, Irving Townsend. I also know she liked the fact that I had employed Cal Lampley, a black assistant, now on the staff at Eastman School of Music. Looking back, it pleases me to think I was the first person in the record business to hire a full-time black A&R man, without assigning him only to black artists."

The discord or artistic unhappiness John Hammond predicted for Mahalia, if she were to take the step into the "big time," did happen to some degree, both in her recording for Columbia and in her TV appearances. He said that at her death, she'd lost a major share of her original black audience. Columbia provided elegant accompaniments, but the earlier drive and looseness of her Apollo recordings were gone.

"I begged them," she said, "to let me record one of my favorite songs, 'God Is So Good to Me,' but they gave me a hard time. The song was so very much a part of my life, like, I thank God for bringing me down in Louisiana to my life today. You know, I'd forget for a moment or two that I was in a studio, standing before a mike, while singing that song. Let me tell you this: There was more of the original Mahalia in the Apollo album 'Bless This House' than in all the other albums I ever made for Columbia!"

And there was the matter of *time* during rehearsals. Mahalia must have lost her temper more than once to have engendered an alleged complaint from someone in the Columbia management: "So, what's the bitch kicking about now?"

"Time is important to me," she explained, "because I want to sing long enough to leave a message. I'm used to singing in churches where nobody would dare stop me until the Lord arrives! But the first thing those television and recording folks would do was to start warning me, 'Look out! Watch your time!' And then, first thing you know, they'd start cutting the song!

"I always got the feeling that some of those producers and studio people were trying to slick me up, turn me into a commercial entertainer. But real gospel singing is not *just* entertainment. They often gave me the feeling that I'm just too sad for their program." And at some of the TV rehearsals, she felt she was a musical interlude between a joke and a dance routine.

Despite the commercial strictures put upon her, by the close of her first year as a Columbia recording artist, disc jockeys from coast to coast were recognizing her voice, her unique style, playing her records with a "Now folks, here's another great Mahalia hit to set you tapping and swaying!" And Mahalia was enjoying every minute of the flattery, even as she complained: "I hate to see how the commercial world takes over the songs that have been the strength of my people. How they turn them into jived-up nightclub acts and rock-and-roll recordings. The dignity of a colored church—religion itself—is being debased, so that a few people can make some fast money out of them."

John Hammond had predicted it all. Mahalia Jackson was too authentic and volatile an artist to have been left only to the black church and lawn parties of her girlhood. Her voice was destined to leap forth and compel the market to make a place for her and those who followed after. She would pay a personal price, but, like the rising civil rights movement, she had now too large a claim for justice and a fair share of the American pie to remain silent, nurse grudges, and not surrender just a little to success in the commercial world.

Mahalia, the "back'a town" little girl from New Orleans, was beginning to make a lot of money. She admired that state of grace almost and as much as the other more unattainable one for which her voice cried out. And thus grew her fame as the civil rights storm approached.

"Blowin' in the Wind"

By the mid-1950s, Mahalia had much on her mind besides career and money. Like most black entertainers, especially those who had grown up in the South, she had suppressed her anger at racial discrimination. Silence was galling but necessary for blacks in order to survive. Non-compliance with codes and statutes of segregation was *sub rosa*. Mahalia blistered at the mockery of the "Separate But Equal" doctrine that the Supreme Court had approved. Nothing had changed in more than fifty years. In the mid-1950s, President Eisenhower, expressing an absurd opinion, confronted the intention of the courts to liberalize the doctrine: "I don't believe you can change the hearts of men with laws or decisions."

To that, Dr. Martin Luther King, Jr., replied: "While it may be true that morality cannot be legislated, behavior can be regulated. The law may not change the heart, but it can restrain the heartless. It will take education and religion to change bad internal attitudes, but legislation and court orders can control their external effects."

Especially upsetting to Mahalia, who had by now ripped open a hole in the segregation scrim, were the weird contradictions that still existed between the enthusiastic white response to her concerts and TV appearances and the more often than not cruel reception still reserved for touring black talent in the South: "The minute I left the concert hall I felt as if I had stepped back into the jungle."

In between concerts and occasional radio appearances, she toured with Mildred Falls and her cousin John Stevens, a young Chicago actor who drove the women in Mahalia's lavender Cadillac into the South from Chicago.

"It was a nightmare! There was no place for us to eat or sleep along the major highways. The restaurants wouldn't serve us. At food stops, teenage girl carhops would come bouncing out to the car and then stop dead in their tracks when they saw we were 'Negroes.' They'd spin around without a word and walk away. Some gasoline stations didn't want to sell us gas or oil. Some told us no restrooms were available. The looks of anger at the sight of us colored folks sitting in a nice car were frightening to see."

In the pre–civil rights days, black touring companies competed against one another in giving concerts to a widely scattered church circuit. It was the custom for performers to take their pay in cash at the conclusion of the program, go out to their cars, and drive off into the night to their next date. Those who were the stars of their little companies got to sleep in their cars if no black hotel or black-owned farm was within driving distance. Non-starring performers slept on the ground, rolled up in blankets. Many were the nights Mahalia had to sleep in her lavender Cadillac. "To find a place to eat and sleep in a colored neighborhood meant losing so much time," she recalled, "that we were finally driving hundreds of extra miles each day, to get to the next city where I was to sing, without stopping, so that we could have a place to eat and sleep. It got so we were living on bags of fresh fruit during the day and driving half the night, and I was so exhausted by the time I was supposed to sing, I was almost dizzy."

Those were the first years of her ascendancy in the white music market, with exulting audiences moving her to tears with their salvos of cheers. She would shake their hands and thank them for their applause, but she was strongly moved to disparage their tributes: "I felt like saying: 'How big does a person have to grow, down in this part of the country, before he's going to stand up and say—Let us stop treating other men and women and children with such cruelty just 'cause they are born colored!'

"From the days of slavery, black men have worked inside white people's homes as pantry men, butlers and chauffeurs, helping bring up little boys and girls, and taking pride in being a trusted part of the family. Up north today many families rely on black maids and baby sitters. So if they trust their most cherished possession—their child—to us, tell me, how can they have such hate and fear of us?

"You find black men who marry white women, but I think they are going the long way around to happiness. I have nothing against intermarriage, except that it means a black man is leaving behind the black woman who has worked and suffered with him since slavery times.

"It's been the black woman in the South who has had to shoulder the burden of strength and dignity in the colored family. Even when she let the white man have his way with her—and it must have happened often because many, many blacks in this country are not black like myself—I believe she went with the plantation master or the field overseer or his sons, so they would be easier on the colored men on her plantation.

"Down in the South, the white man never gave the black man a job he could be proud of. He always called him 'Boy' or by some nickname. It was hard for colored children to be proud of fathers who were treated like that, and it was usually the black mother who had to keep a certain dignity in the family to make up for the inferiority the white man inflicted on her husband.

"The plain common black church woman has always been the backbone of the colored church and has given the last dollar she earned scrubbing floors, to the collection plate.

"Today, fifty percent of black business, North and South, is run by black women. They have made beauty culture a big business. Black women, like Madame C. J. Walker, own cosmetic shops from coast to coast, and every black community has its hat shops and flower shops and dancing schools for children. So, if a black man wants to marry a woman he can be proud of, there's no need for him to look around for a white woman."

Mahalia admitted that most of the above dialogue had been only *thoughts* in her head, as much as she wanted to shock her white audiences, the admirers of her art, with her feelings. But she controlled herself, she said, as she smiled her way through crowds of white audiences; she never challenged the painful inequities in the society that honored her art. After all, it was the time *before* Little Rock and Montgomery, she was quick to remind one.

Rosa Parks, in 1955, was a forty-three-year-old seamstress living in Montgomery, Alabama. Until the evening of December 1, she was a non-political woman. But this evening, on her way home, she was especially tired. She boarded a bus and took the first available seat in the front section that, under the city's laws of segregation, was reserved for whites only. A white man entered the bus behind her, and the operator ordered her to get up and give him her seat. Rosa Parks refused, and by such an outrage brought the force of the city and law down upon her. She was arrested, briefly jailed, and ordered to stand trial on the grounds of having violated the segregation laws of the city and state of Alabama.

In the fall of 1956, Mahalia had an unexpected visitor from Montgomery, Alabama, the Reverend Ralph D. Abernathy, a colleague of Dr. King's and an active leader in the growing civil rights movement. Dr. Abernathy had come to ask a favor of Mahalia. He explained that the St. John A.M.E. Church of Montgomery intended to honor Mrs. Parks and the year-old black boycott that had catastrophically disrupted city life. He told her that he was organizing a seminar on non-violent protesting and the politics of social change, and wanted her on the program to provide a bracing musical interlude of her own choosing. Would she come and what was her fee?

Dr. Abernathy was to report later that Mahalia answered in her purposely cynical-black rhythm and grammar: "Yes, but, I ain't comin' to Montgomery to make no money off them walkin' folks!"

Montgomery would never be the same again. On Thursday, December 6, people began to pack the church by noon. By nightfall, stragglers couldn't force their way into the basement where the concert was to take place. It was, literally, a packed house when Mahalia arrived on stage. She was made aware of the fire regulations for the church, and there was a discussion concerned with the threat of a riot. With that worry in mind—overcrowding and riot—she opened with what she called a "soft number": "I've Heard of a City Called Heaven." She sang with her eyes shut tight.

After the thunderous applause died down, she began gently with "All to Jesus, I Surrender . . . He's my blessed Savior." Working her audience into a bounce, she rocked them with "I'm Made Over . . . I'm going to sing and never get tired." The rafters rang with shouts: "Sing the song, girl!" and "Oh, yeah, God, oh yes, Jesus!"

Her program continued into the night, ending a day of fasting and thanksgiving. She swung into two favorites, as the celebrants leapt to their feet in exultation: "God Is So Good to Me . . . and I don't deserve all of His good." Then came her trademark: "Move On Up a Little Higher."

The church rocked and shook from stem to stern, floor to ceiling. To calm the emotional waters, Mahalia concluded her pre-Christmas program with the tender "Silent Night, Holy Night," one of her giant hits. And only her voice, with its amazing natural artistry and passion, could make that simple song of everyone's childhood memory come out sounding as broad and uplifting as a Bach B Minor Mass.

In early 1957, having accumulated an impressive nest egg, Mahalia bought a handsome, single-level house and garage at 8353 Indiana Avenue, a predominantly white neighborhood in Chicago's South Side. She paid forty thousand dollars for it. A few weeks after she moved in, an unidentified person, allegedly an irate neighbor, fired air-rifle pellets into her living-room windows—an unequivocal warning to a prominent black daring to invade white turf.

For many days, Chicago police patrolled her property and the neighborhood, but never during that stressful, ominous time did Mahalia even consider moving out of the community. News of the attack on her house had made headlines throughout the country and naturally came to the attention of the legendary journalist Edward R. Murrow, whose CBS weekly telecast, "Person to Person," was the most watched news show of the fifties. It was Murrow who had originated the format of interviewing famous people "at home," in their own environments, a format to be copied in ensuing decades by David Frost, Barbara Walters, and vulgarized completely by the through-the-keyhole-cum-camera of the "Rich and Famous" shows.

Edward R. Murrow, famed for his forthright, pithy interviews, took advantage of the shocking assault to Mahalia Jackson's house, if not her person, and paid her one of his famed person-to-person visits. It was a great opportunity for the journalist with a conscience. Murrow charged that the threat to Mahalia was an incident in the rising tide of racism. And whether the audience liked to hear such news or not, he warned his listeners that, if a celebrated black artist could be physically threatened to leave a middle-class neighborhood of Chicago's South Side, then, surely, no black person, rich or poor, might expect to be secure anywhere outside the designated ghettos of the nation.

The press had estimated the crowds milling around Mahalia's house for more than three days to be as many as 3000, drawn there by the news of the attack, the celebrities who came to commiserate and show their solidarity with her, and the "Person to Person" telecast. Years later she remembered it all as having a happy ending: "A group of people came to my door and asked if there was anything they could do to help. I said I would love to have all the children they could find. I wanted them to be with me on the TV show. Their mothers dressed them up, white and colored children, and they all came, and we had a wonderful time. I cooked up a Lousiana Everything Gumbo—red beans and rice, ham and shrimp. We all ate together in my crowded kitchen—the cameramen, the children, neighbors, just everybody!"

That was the last and only time anybody ever took a potshot at Mahalia Jackson.

Dinah Shore, one of the top TV personalities of the day, was the first white star to insist that her network, CBS, enter into a contract with Mahalia for an appearance on the "Dinah Shore Show."

It was to be an historic alliance—the blond talk-show hostess, TV's darling, and the sable-skinned empress of gospel—that would make a big dent in the rigid Jim Crow hiring practices of the entertainment industry.

The two women, both powerhouses in their own style and manner, had privately admired each other's public performances, but their studio meeting in Hollywood was a cautious surveillance of each other. Would their personalities collide and make a collaboration impossible? Despite the collective fears of their agents, producers, and respective entourages, it was mutual attraction at first sight.

In rehearsal, Mahalia interposed herself regarding the design of the show when it came to her own involvement with it. She said that she would do her standard performance in collaboration with Mildred Falls at the piano, as usual. Mildred would set the beat at the piano, and the orchestra was to join them along the way. Mahalia insisted that the orchestra take its cue from Mildred's beat and tempo. There were raised eyebrows and maybe more than a few open mouths in the production department, but Mahalia had her own way. The result, the record says, was a smashingly swinging Mahalia favorite that was talked about for days after: "He's Got the Whole World in His Hands."

With the guidance of cue cards, Dinah and Mahalia debated the merits of the blues, with Dinah assuming that Mahalia's whipsaw style contained blues elements in it. Mahalia quickly disabused Dinah of that notion. Mahalia said that she loved the blues, to listen to, never to sing; she spoke of Bessie Smith and how Bessie's blues style was unequaled, "the greatest." Dinah then explained that as a child she had been exposed to "white" blues in the church her parents attended. Prudently, Mahalia suggested that Dinah must truly know the difference between the blues and gospel.

Dinah, grinning, agreed that she did, but suggested that Mahalia explain her view of it.

Mahalia responded: "The blues, baby, is when you're feelin' low . . . when you're down in the mouth. But gospel is always happy, a joyful sound. You *know* when you're up an' feelin' good!"

The two women then sang a rip-roaring duet of the great antiwar gospel, "Down by the Riverside." The next day the critical reviews were ecstatic. It had never happened before in just that way, over the airwaves—two stars in a happy conjunction of black and white.

Mahalia was pleased with her TV reception. It confirmed her own belief that more contracts and hefty fees would flow from such a headline program. She was right. Like a torrent, the offers rushed into her agents' offices: New recording dates with Columbia Records, numerous high-flying TV guest shots, and a clutch of nightclub offers, much to Mahalia's amusement and secretly pleasing her conceit, tempting the devil himself.

In the following months, her agents at William Morris orchestrated a brew of TV spectaculars for her, starring such blockbuster talent as Bing Crosby, Perry Como, Steve Allen, Red Skelton, Ed Sullivan, Milton Berle, Danny Kaye, Jimmy Durante, and Dinah Shore again.

In 1958 came the film role in Universal's remake of *Imitation of Life*. Though Mahalia was no actress, studio executives were prepared to gamble on her ability to bring in big audiences to the box office.

Directed by Douglas Sirk, the film was never more than a mediocre melodrama, even though Lana Turner, at the top of her trade, was cast in the lead. The studio was gambling on something else to help make the film a box-office smash, something, in Hollywood fashion, reprehensible, venal, and shadowy. The studio was banking on the sordid pre-release headlines concerning the private tragedy in the Turner household. Shortly before the film was cast, Lana's daughter Cheryl had murdered her mother's lover Johnny Stompanato, a small-time gangster. Casting Lana Turner as the emotionally charged mother-actress who had to confront the painful adventures of her troubled children could only add to the exploitation value of the film, or so went the reasoning.

Mahalia's role was that of a Louise Beaver update, a segregation-era caricature of the happy "colored" servant. For the big, pivotal scene, the studio hired a Hollywood church and added a mob of screaming extras. Mahalia was persuasive when she sang, delivering a robust rendition of the 23rd Psalm from the pulpit. In the funeral scene she cried, magnificent and voluptuous, on cue.

I never did get the chance to ask her whether she thought that the Hollywood money justified her ill-advised performance in the movie.

Soon after the film's release, she made a guest appearance on TV with Bing Crosby and Dean Martin. There was plenty of bemused badinage with the masters of the flip, hip comedy routine. Mahalia was now definitely in the big time, the big money.

But the dubious trade-off, the price for a black artist on TV and the screen before the civil rights movement compelled non-racial characterization, was the sometimes repellant, abusive caricature, as in the case of Mahalia being costumed and directed as a rocking-chair, bandana-head 'Mammy' singing "Summertime" on a farm porch set, or a spiritual like "Sometimes I Feel Like a Motherless Child."

Interlude
for
Mildred

From the start of her career in Chicago, in the early fifties and on, Mahalia had employed many talented pianists and organists to accompany her, beginning first in her church work and later for recordings, radio, and television. But of all the talented musicians she had worked with, none was the equal of Mildred Falls, a kind of genius in her own right when it came to using the keyboard to its fullest, as a percussive as well as melodic instrument.

It was only Mildred who never lost her way in the call-and-response gospel style that Mahalia had made her signature. Without Mildred's blues chords, her triplets and four-

four irresistible, inimitable bounce pushing Mahalia's voice on like a jockey's whip, Mahalia's art would have been the poorer.

Whether accompanying on the piano or organ, Mildred's commanding music had the resonance of choir and congregation. She gave Mahalia the latitude she needed, the freedom to ad lib new lyrics, break time, alter the melodic line in the heat and passion of building the meaning of a song, all the while steering the singer onward in her unrelenting gospel mission. Theirs was a collaboration of equals, even if Mahalia's persona and art were the theatrical attraction.

Mildred adored her and cherished their alliance. But Brother John Sellers, who had been treated as "son" by Mahalia at various times in their lives, has a memory of the relationship of the two women that is far from flattering to his "mother."

"Mahalia gave Mildred whatever *she* wanted to give her. But Mildred really was everything, she was fantastic!

"Mahalia's reputation was spreading fast when you first met her," he reminded me. "And Mildred was pleased to get so much work regularly. But then, Mahalia was the type of woman who wouldn't pay anybody if she could keep from doing it. That's how she got rich and kept her money," he added in a voice filled with asperity.

Recognition for Mildred Falls's talent in her own right is long overdue. She died in Chicago in the 1970s, poor and rejected by the one person who should not have abandoned her. Mildred's unique contribution to Mahalia's success ought to have been decently compensated, but the opposite became her reality. The truth was, Mildred was abused by being abysmally underpaid, and, when her services were no longer desired (when she had the nerve to ask for a raise), she was made a pariah and conveniently forgotten, curiously not only by

Mahalia but by everyone else in the Jackson circle. There was shame and perplexity in Brother John's look and tone, as he thought back on the fate of Mildred Falls.

When Mahalia moved into television, it was her practice to instruct the producers to send her accompanist's check to her, the stated amount arranged by Mahalia. What Mildred then received was whatever Mahalia wanted to give her. It was a time when Mahalia was earning $3000 a night, frequently as much as $7000. Mildred was being paid $200 a week and expenses, which were not more than $100 when they stayed in a hotel and ate in restaurants. When Mildred asked for an additional $100 a week, Mahalia summarily fired her.

It had been Mahalia's way since childhood to keep every penny she made as close to her as possible, literally, to leave her job with the cash in her bra. The part of Mahalia that had remained primitive, wary, and poor, no matter the thousands pouring in, hated certain aspects of her success such as those times when she had to allow her William Morris agents to pick up her check and later pass it on to her, *minus* their 10 percent commission.

Brother John recalls that she would never take criticism of her behavior from anyone. "Not from Mildred, or me, or any of her husbands. She had the habit of sayin': 'I'm Mahalia Jackson, you hear?!'"

In the mid-1960s, when Mildred Falls was down on her luck, sick and on home relief in Chicago, no one ever came to her aid. "We didn't do right by her," Brother John commented. "But you couldn't talk to Mahalia about Mildred's situation. She didn't want to hear about her. When Mahalia had money, nobody could talk to her. She used to close out anything she didn't want to have anything to do with."

As Brother John grew older his early adoration of Mahalia shrank in size. He had evolved into an artist in his own right—a blues and gospel singer of note, particularly in Europe—and she had begun to reveal aspects of her character that distressed not only him but other friends and colleagues as she made her commercial ascent and success. He recalls the times he had traveled with her as a young boy, from one church concert to another, after she had left the Johnson Singers. They would move about in buses, staying overnight in the homes of fans and devotees. She was earning between five and eight hundred dollars for a nightly performance and gave him whatever sum she wanted, and when she pleased. He was expected to be grateful. "If we had four or five dates, Mahalia would dig into her purse and say to me: 'Here, boy, there's fifty dollars. That's yours now!' Just like that. And, you know, in the end, that's the way she treated Mildred Falls."

One of the regrets of his life was that he didn't have the courage or maturity to move away from Mahalia's influence and temper, that he didn't separate himself from the "queen's circle" that had closed against Mildred. For she was the best.

New
Roads

It was very good news for Mahalia when George Avakian asked her to participate in a New York concert that he and his wife would produce in the celebrated Town Hall, noted for its illustrious *ragoût* of musical debuts, historic lectures and small theatrical events. Though with none of the bravura surrounding Carnegie Hall, it was, in its own right, an important New York stage. She was to be cast opposite the world-famous French baritone, Martial Singer, whose program consisted of a variety of French and German art songs. The Singer-Jackson concert, the third produced by the Avakians in the 1957 season, was entitled "Variations on the Folk Theme."

Her appearance at Town Hall, according to Avakian, drew a larger audience than anticipated, but with Martial Singer it was hardly a rousing box-office hit. "There was a fair representation of blacks in the audience, but the larger crowd we had anticipated may have been put off by the fact that it took place at Town Hall, off the track for her public not attracted by a European white singer doing art songs of European folk origins. Yet, the audience was so responsive, they resisted Mahalia's finishing her part of the program—she had to go into overtime, singing encore after encore. I remember, because I got billed for it!"

Avakian is certain that he may have been the only impresario who ever paid Mahalia by check. Shortly after the concert he got a call from her that she had lost his check. He proposed that they wait a few weeks to see whether it would turn up. The disappearance of the check became a thorn in his side. Suggesting that perhaps, inadvertently, she had stuck it into one of her Bibles, he would call her each month to remind her to make a search for it. Finally, he sent her a replacement check and asked her to tear up the first one should she ever find it. To this day, he is convinced that somewhere in Chicagoland there exists a unique, uncashed check made payable to "Mahalia Jackson," because she had wanted nothing to do with that kind of paper.

As gospel songs and spirituals were considered serious and sacred music for black church congregations, the "blues" were secular and often viewed as sinful, what with their origins in southern streets, back alleys, and the black cabarets called "barrelhouses." Brother John Sellers, growing up in Mississippi river towns, was exposed to both gospel and blues, and eventually jazz: all the natural ingredients of black music.

By the end of the 1950s, Brother John, then in his mid-twenties, made hard decisions about his life. He admits, forty years later, that he had become unhinged from the stress of being bonded to Mahalia and her zealous chase after her own fame and fortune, "her destiny." He had begun to struggle with his own sense of destiny, for by then he had proof enough of his own musical talent and could, doubtless, make a living with it; he didn't have to depend upon handouts from Mahalia anymore. It was time to test his own wings in the music world, and the opportunity, the incentive, came from his friendship with Chicago's great blues artist, Big Bill Broonzy.

Big Bill understood Brother John's need to cut his dependency chain with Mahalia Jackson. As she became aware of Brother John's plans, Mahalia resisted and attacked him for his failure to remain loyal to her needs.

Big Bill came north to Chicago in 1937 from an Arkansas river-bottom farm. He'd grown up driving a pair of mules pulling a hoe. As he worked the land, he composed blues songs, primitive and beautiful, which he put to memory until he found someone to record them. On one of John Hammond's southern talent searches, he met Big Bill, adored his art, and paid his way to Chicago, where he quickly became the stellar attraction at the Blue Note cabaret. Soon after, Hammond, midwife to the best of America's folk artists, signed Broonzy for his Christmas concerts at Carnegie Hall in 1938 and '39, "From Spirituals to Swing."

Broonzy took his first, sophisticated, New York audiences by storm. Backed up by three of the most famous boogie-woogie pianists of the day—Meade Lux Lewis, Albert Ammons, and Pete Johnson—Big Bill strummed his guitar, growled his gutbucket black-blues and brought it to 57th Street with a musical vengeance never to be forgotten. In between his numbers he created a rapt silence in the audiences by sharing a dream he'd had that found him sitting in FDR's presidential chair behind the desk in the Oval Office of the White House, playing President!

By 1953, John Hammond had been drafted by Maynard and Seymour Solomon, the producers of Vanguard Records, a small company that specialized in the classics, to develop a line of jazz recordings. Among the stellar artists of the day, he chose Brother John to record two albums before John took off for concertizing in Canada and Europe. The first album, "Brother John Sellers Sings Blues and Folk Songs," was produced with Sonny Terry on the harmonica, Sir Charles Thompson, piano, Jo Jones, drums, Walter Page, bass, and Johnny Johns and Freddie Greene, guitars. A couple of months later, Vanguard, with Hammond as producer, put the trio of Sellers, Sonny Terry, and Johnny Johns together to do "Jack of Diamonds and Other Folk Songs and Blues." The results for critics and market alike were exceptional, and provided Brother John with a singularly potent passport to the European marketplace.

Big Bill, watching young Brother John "come up," decided that he had stardom in his soul, and when the older artist made his plans to leave the South Side of Chicago for places like London's Royal Albert Hall, he gave John an open-ended invitation to join him when John felt ready for the leap and separation from Mahalia. At first John tested the waters alone, out of the States, in the cabarets of Quebec. It was there that he felt the first flush of success, a sense of independence from "Mama" Jackson. By 1958 John must have felt *ready,* for it was then that John accepted Big Bill's invitation to join him, and the two conquered England, Paris, and Brussels together, with John, for a time, putting Chicago and Mahalia Jackson behind him while he performed and made historic recordings with Big Bill Broonzy.

Mahalia, in the meanwhile, reasoned to herself and anyone else who would listen: Who the hell was John without her? He could do as he damn well pleased, live and work with Big Bill, if that's what he wanted; she could easily dispense with her protégé for a year, without any pain whatsoever!

Anyway, Mahalia didn't have pleasant memories of London. In 1955, when Big Bill had achieved celebrity status in England, generous man that he was, he was in a position to introduce her to the great audiences in the Albert Hall. He offered to make room for her on his program of blues and jazz, and she accepted. But London's first response to her gospel program was as cold as the bitter winter weather outside. The audience that filled the Hall was uptight, conservative, and bourgeois. The reporters of the event wrote that Mahalia Jackson seemed to be in a subdued mood during her entire performance, and not her typically inspired self that they had come to expect from her records. The audience, sitting on their hands, dignified, rigid, and regal, there to be seen more than to receive, refused to allow Mahalia to make her own "time." Her slow, dark, passionate build to the religious light and catharsis, her infectious gospel beat fell on deaf ears.

Another reason for Mahalia's negative London reception, aside from the stiff-backed Albert Hall audience, was the fact that Mahalia had been ill in Chicago a few weeks before she accepted Big Bill's offer to join his group there. It was reported that her performance was tinged with melancholy. One critic wrote that she had lost much of her "magnificence" except for one bluesy song (not otherwise identified) that was indeed poignant as she sang the words "I had a friend." It reminded the critic of the harsh defiance of Bessie Smith at her best. Mahalia, understandably or not, wasn't appreciative of Big Bill Broonzy's gesture, and she complained that it was wrong, stupid even, to have introduced her and the gospel on his program. She left London in a huff and went on to Scandinavia, where the applause, happily, was more substantial and accepting, which again might have been for multiple reasons. She was feeling better, and the audiences were looser, more tuned in, working-class, hip, and eager for the new, black sound and its contralto empress.

The Beatles were still an obscure group working their way across Europe in 1958, while in the States "the blues" and "folk" had taken over completely. Brother John now returned to New York after his successful sojourn abroad, to help launch a new nightclub in Greenwich Village, called Gerde's Folk City.

Between the Korean and Vietnam wars, students began to hang out in small, inexpensive nightclubs where drinks were cheap and the entertainment was music in search of roots: folk, jazz, and the ubiquitous blues. In San Francisco it was the Hungry Eye; in Chicago it was the Blue Note. In New York the place to "be" was Folk City, whose patron saint was the dying Woody Guthrie. The young artists who performed in those clubs, strumming their guitars, singing old folk, new folk, old blues and new blues, would become the immortals of the pop music world of the 1960s, the Vietnam War, the Civil Rights movement; their words and music would be the tender conscience of a generation daring to question the nation's political drift between Hiroshima and the Vietnam debauchery.

For a few seasons Brother John performed regularly at Folk City and emceed most of the shows. It was an exceptional time for young musicians with a message: They went from anonymity to commercial stardom in the blink of a decade. Mary Travers tested her talent at Folk City before anyone dreamed of the trio, Peter, Paul, and Mary. Joan Baez, while still a college student waiting on tables in Boston, captured her first New York audience in tryouts at Folk City. John Hammond made it one of his favorite hangouts. It was there that he came upon a young man from Minnesota, calling himself Bob Dylan, and signed him to a contract. Paul Simon and Art Garfunkel got their early exposure in front of the aficionados at Folk City. All were to become the stars, some of them the social critics, of an era.

It was a few years later that Brother John joined with the dancer-choreographer Alvin Ailey in the creation of two folk ballets that became the most heralded works of the Alvin Ailey Dance Company: "Blues Suite" and "Revelations."

John arranged and sang the score for "Blues Suite," using the songs of the backwater and Depression blues of the South, including two of his own compositions: "Me and Old 'Frisco" and "Something Strange Is Going on Wrong." For the ballet "Revelations" he used the literature of gospel church music and guided Ailey musically in the building and translating the fervor and excitement of the gospel experience—which John knew so intimately—into the language of dance. Together they made dance history by placing black, southern folkways into the mainstream of world ballet theater.

A few seasons later, Brother John went to Europe and the Far East to perform and headline on his own. His raw, throaty, elemental blues style was an ambassador for the laughter and tears, the sensual power of American black country music. Through such performers the world could begin to understand something of the dark side of the American South and its biracial fabric.

It *had* come to pass that Mahalia would be forced to overcome some of her resistance in crediting Brother John with a talent all his own. She was heard to say, about the blues music that she, herself, would never perform: "When black people stop singing the blues, then there'll be no more nothin'! Because the blues has made, is, American music. I *know* the blues and the gospel, and they will still be around when all the rock and stuff has gone. The blues is always around."

Violence
in America

The spectacular concert Mahalia had given in support of the black resistance in Montgomery, Alabama, brought her into contact with Dr. Martin Luther King, Jr., and his wife, Coretta. Mahalia was frequently a guest in their home and shared their fears of a political nightmare to come.

Mahalia clearly saw herself as a player in any future black resistance movement. Hadn't she resisted humiliation and injustice her whole life? From the "back'a town" days of the Louisiana ghetto, even to the time when she had become famous touring the country, she still felt the injustices meted out to blacks, famous or not. Money and success eased the way but didn't obliterate that swallowing of anger and outrage felt by every person of color.

And out of the turbulence of the Montgomery bus boycott came Dr. King's spiritual leadership, one of the watersheds of American history, the non-violent, passive-resistance, political movement among blacks and allied whites.

Dr. King and his associates in the Southern Christian Leadership Conference were certain their ideals of political justice and economic freedom would soon erupt into a national crusade. Mahalia, listening to the dialogues of her churchmen—Dr. King, Rev. Abernathy, and their supporters—saw her vocal art as an ally for the minister leadership, her voice as a weapon for change. In time, her covenant with Dr. King and the SCLC would bring her, inevitably, into conflict with her lifelong connection to the Chicago-based, cautious, if not reactionary, authorities of the Baptist church.

At high noon on May 17, 1957, the Southern Christian Leadership Conference held a Prayer Pilgrimage for Freedom at the Lincoln Memorial in Washington, D.C. There were fourteen speakers on the program and Mahalia sang "I Been 'Buked and I Been Scorned."

The last speaker was Dr. King. He stirred the thousands in the predominantly black audience that stretched out before him, and also the nation, as he urged his listeners to do the miraculous: abandon wrath and malice, and ask God to turn the hearts of men and women toward one another. He insisted that it was now the time for economic and social inequities carried over from slavery days to be jettisoned, that the spirit of brotherhood prevail. As the crowds cheered, stamped their feet, and waved thousands of white handkerchiefs, King paused and looked back at the majestic figure of the contemplative Lincoln. Then he turned to face the audience and charged the government with betrayal of the people. He waved his arms above his head and shouted: "Give us the ballot now!"

Democrats and Republicans alike, senators and congressmen had to have reacted to his message that day. He had thrown down the gauntlet to the people's alleged representatives. Black America would reach *beyond* compromise and claims for restitution for the unpaid work of slavery.

The time for compromise was over.

Dr. King made serious, even dangerous inferences concerning the course he was encouraging the SCLC to take in any anticipated struggle. He now insisted they acquire simple self-defensive tactics—a strategy that made allies of many previously neutral blacks and whites.

To a group of journalists who were pressing him for a definition of the goals of the SCLC, Dr. King said: "There is more power in socially organized masses on the march, than there is in guns in the hands of a few desperate men. Our enemies would prefer to deal with a small armed group rather than with a huge unarmed, but resolute mass of people. It is necessary that the mass-action method be persistent and unyielding. The determined movement of people incessantly demanding their rights always disintegrates the old order.

"Our present urgent necessity is to cease our internal fighting and turn outward to the enemy, using every form of mass-action yet known . . . create new forms . . . and resolve never to let them rest."

On May 15, 1954, the Supreme Court drew a full house. The case on the docket known as *Brown et al. v. United States* dealt with a reconsideration of the Fourteenth Amendment to the Constitution on the question: Does segregation of children in public schools, solely on the basis of race, deprive black children of equal educational opportunities?

An historic decision came down: the fifty-year-old ruling known as "separate but equal" was now reversed. Race segregation in the United States, literally since the Declaration of Independence, was now legally declared unequal, unfair, and undemocratic.

The outcome was distressing. The collective South now refused to accept (and honor) the Court's historic order. The decision opened the sluice gates of a violent retaliation as in community after community, two separate mentalities confronted each other, identified by many as "the second American Revolution." Segments of white America were preparing to reject a redefinition of civil rights under the Fourteenth Amendment.

Between 1956 and 1962, the politics of rabid racism challenged the constitutional rights of some American citizens. Yet in the White House, President Eisenhower was little stirred by unfolding events. He was strangely neutral as a mob spirit swept across state lines, especially in the traditional South where non-compliance put many cities and towns under the control of the worst segments of the citizenry. Very little moral leadership was to originate in the White House as more than 50,000 national guardsmen were called out to police a hundred cities.

The governor of Arkansas seemed to encourage the formation of lynch mobs that ultimately confronted the National Guard, directed to escort black students into the high school of Little Rock. The racist and self-righteous Senator Richard Russell of Georgia wired Eisenhower that the guardsmen assembled at Little Rock were no better than an army of Hitler's storm troopers, and that their presence there was an evasion of states' rights.

With the assumption of the Kennedy administration, an attitudinal change took place in the President's office and within the Justice Department as the "revolution" progressed—more youthful and caring, but those in power were by no means ready to act with the compassion and dispatch for which the times cried out.

Race relations collapsed in Birmingham, Alabama, under the violent police control of Commissioner Eugene "Bull" Connor, who used snarling police dogs and fire hoses against a thousand protesting black students, orchestrating a political scene of hell that became emblazoned in the nation's psyche.

A protest march by the three principal black leaders—Dr. King, Rev. Fred Shuttlesworth, and Rev. Ralph Abernathy—began the morning of Good Friday, April 12, 1963. The marching reverends singing hymns, with a hundred and fifty demonstrators, were confronted by "Bull" Connor and his troops. The ministers were arrested as leaders of the protest march and jailed. For Dr. King, this was his thirteenth arrest in the South since beginning the anti-segregation movement. More demonstrations continued through the month.

In Jackson, Mississippi, Medgar Evers, secretary of the state's National Association for the Advancement of Colored People and an activist in opening the state university to black students, was shot by an assassin's bullet from a high-powered rifle as he was about to enter his home.

Hodding Carter, editor of the Mississippi *Delta Democrat Times* at Greenville was quoted as saying: "Let it be said, hopefully, that the struggle between moderates and extremists of both races will eventually be resolved in favor of the moderates. If it does not, God help us all!"

On a quiet Sunday morning in a children's Bible class in a Birmingham church, September 15, 1963, a bomb exploded, killing four little black girls and seriously wounding fourteen other children. Their Sunday Bible lesson was entitled "The Love That Forgives."

Four U.S. senators —Gary Hart, Thomas Kuchel, Hubert Humphrey, and Jacob Javits—framed a resolution for their fellow senators, proposing that the following Sunday be set aside as a national day of mourning in memory of the four murdered little girls.

Dr. King and six fellow ministers went to the White House to discover that President Kennedy had already appointed a two-man, white committee to represent him at the memorial services in Birmingham. Wrote I. F. Stone, an independent Washington journalist: "Surely, he could have appointed at least one of the two black judges from the Court of Appeals, to dignify a mission of mediation. He might have insisted, for once, after so terrible a crime, on seeing white and black leaders together. It is as if, even in the White House, there are equal, but separate facilities."

The summer of 1964 was a nationally decisive one, because of a few hundred northern college students who used their summer vacations to live in the South and serve the black voter registration program under the auspices of the Student Non-Violent Coordinating Committee.

In mid-June, after the start of the project throughout the state of Mississippi, three college youths disappeared near the town of Philadelphia, following their arrest on a trumped-up charge of speeding. Six weeks later, after a widespread coverup and collusion by local citizens and police, FBI agents found the bodies of Michael Schwerner and Andrew Goodman, white students from New York, and James Chaney, a local black student, buried in an earthen dam on the outskirts of the town. They had been executed Klan style.

Twenty-four years after that satanic act of vengeance, in

1988, British film director Alan Parker produced *Mississippi Burning,* a film that centers on the Philadelphia murders, but contrives a plot that makes heroes of the two FBI agents, rather than the victims. Mahalia Jackson's recording of the gospel song, "Take My Hand, Precious Lord," is the opening theme of the soundtrack under the main credits of the film. She would have been pleased that her voice, twenty years after her death, was used by a new generation to reaffirm her war against bigotry.

The unrest in the black/white body politic moved steadily onward toward an apocalypse. On the evening of February 21, 1965, the black nationalist leader Malcolm X (Malcolm Little), while speaking at the Audubon Ballroom in Harlem, was shot point-blank by three men who had begun an argument in the audience, obviously as a cover for the assassination.

Malcolm X, a short time before, had broken a long relationship with the top Muslim spiritual leader in the U.S., Elijah Muhammad. The police suspected that the killers were hirelings of Muhammad. That same night, the Muslim mosque in New York was fire-bombed. It was suspected that the protégés of Malcolm X had struck a retaliatory blow. James Farmer, a leader of CORE (Congress for Racial Equality), had a different suspicion; he believed it was the agents of the Mafia opposed to Malcolm's crusade against narcotics and crime in the slums who had orchestrated his murder.

The Kennedy brothers, for some time, had warned the power structure of the South that it faced a choice between the non-violent philosophy of Martin Luther King, Jr., or the belligerence of the Muslims. Even as the four black children were bombed to death in the Birmingham church and the city was ripped apart by the black outrage and response, there was little evidence that presidential advice was to be influential in changing the convictions of southern whites.

Though Mahalia moved through an eclectic world of race, politics, and commerce, there is no indication that she provided any support to the Black Muslim Movement. To the contrary: A Baptist, she had strictly avoided Muslim activities, especially in Chicago, as Malcolm X (long crusading against ghetto rackets by organized crime) began his denunciation of the strategies of Dr. King before his black audiences at the time of the Birmingham bombing: "You need somebody who is going to fight! You don't need any kneeling-in or crawling-in!"

From its beginning, the Kennedy administration had applied a policy, generally, of benign neglect that allowed civil rights workers to be abused or beaten while FBI agents stood by taking notes and photographs. In some instances, agents were aware in advance of the Klan's planning, but rarely provided civil rights workers with any forewarning.

When Dr. King and the SCLC decided to make plans for a second March on Washington, J. Edgar Hoover, it is reported, conceded that the civil rights movement (which he despised) would not wither away or be put down by local or regional white resistance. He now sought other means of containing the freedom movement within legal bounds.

To that *sub rosa* end, King became the personal target of Hoover's resistance to the new law of the land. Unfortunately for the record concerning this personal vendetta, there were more whites than it is comfortable to admit who were of the same vindictive mind as J. Edgar Hoover. And equally unpleasant to admit, the same record shows that Robert Kennedy, the President's brother and Attorney General, did approve Hoover's request to secretly wiretap King's home and office, as well as to allow so-called "fishing expeditions" to obtain data on Rev. King's alleged sexual diversions, womanizing, and political associates.

It is believed that the FBI composed fake letters addressed to the husbands and male friends of the women King was supposed to have had affairs with. But these tactics apparently did not prove effective in turning off the followers who revered King's militant yet non-violent leadership, caring not a tinker's damn for the underhanded machinations of J. Edgar Hoover.

Mahalia painfully recalled how shocked she had been to learn of how many blacks ("They ought to have known better!") resisted giving their support to the nationwide preparations for the second March on Washington, whose chief organizer was A. Philip Randolph, founder of the Brotherhood of Sleeping Car Porters, and his young protégé Bayard Rustin, a labor intellectual.

"Too many people were dragging their feet," Mahalia affirmed. Lots of well-meaning black folks, according to her, had warned Randolph that the event wouldn't and shouldn't come off. The argument was that it was unlikely that people from the West Coast and the Deep South would leave their jobs, especially in the awful heat of summer, to make a second trip to Washington. Anyway, there were few public places for blacks to stay in the Washington area, she was quick to add. Many felt that Congress just didn't give a damn and hoped the event would fail, and blacks and their straggly white allies would be laughing-stocks. And there were those who pointed out that newspaper editorials commenting on the anticipated march were tending to pass it off as "black desperation." People were implying, she said, that it was another wasteful opportunity for "the negroes" to draw attention to themselves, to make trouble for everybody. And as for Dr. King, some said: "Everybody knew he was out to stir things up!"

There was fierce pressure brought to bear on Randolph and Rustin to call off the second march, but they would not be moved.

Mahalia pointed out that, as white Washingtonians began to fear there would be rioting in the streets, many closed their houses and took off for quieter places: "And no wonder. During the springtime demonstrations, men, women and children had been sent to jail by the thousands. They were beaten up, bombed out, and blackjacked by white people, and the guilt of the white people for what they had done and allowed to be done was strong in the South, as well as in the North.

"Driving into Washington the night before the march, it looked to me like we were entering a city about to be captured by an enemy army. Houses were dark and the streets were deserted. There were so many police patrols and soldiers around, you'd have thought the Russians were coming."

What if they should fail? What if the people stay away? Mahalia was filled with a gnawing anxiety. Now that she was close to Dr. King and his associates, she couldn't help but feel concern that they might have overestimated the projected turnout. Could their expectations of building a lasting, radicalized movement of blacks and whites have been premature? She wasn't at the center of planning, but she had begun to feel that her participation was important in attracting non-activist, ordinary folks to Dr. King's cause—now her cause too.

As Mahalia tells it, the evening before the march one of the committee fellows came to her room with good news. "Are they coming!" he shouted. "Mahalia, you're going to see a sight tomorrow like you never saw before. They're rolling into Washington all day and night. Five hundred busloads from New York! Eighteen special trains from the South and Midwest! They're coming in by plane from New Orleans and California! They're coming and coming and coming!"

She cried with happiness, she said.

The next morning, August 28, 1963, the count of arrivals was immense. Since dawn, the streets had been filling up, and even the weather was fooling every pundit who had predicted a traditional summer "hot and sticky" Washington. It couldn't have been more fortuitous—sunny, blue skies, crisp and breezy. By noon, the wide grassy lawns with elm trees that stretched from the White House to the Washington Monument were swarming with celebrating humanity.

Police patrols and military cadres were on a strict riot alert, but there was no fighting or even arguing as the masses gathered. The expected "rioting crowds" never happened. Two hundred thousand people—a sea of black and white faces—were assembling, coming together in cordial proximity and joyful peace. Mahalia said that she remembered the day as the coming of Jubilee, that she didn't know whether to laugh or cry, so she did both simultaneously.

"It was as if the human race had taken a day off from being mean to each other," was the way she expressed it. For the people were arriving in a vast, flowing mass, and singing spontaneously, the spirituals and hymns and the new freedom songs they knew. By noon, an unorganized, impulsive parade of countless thousands waving signs and flags moved in parallel processions along the sides of the great pools, toward the Lincoln Memorial. Mahalia spoke of it as "a nation of people marching together . . . like the vision of Moses, that the children of Israel would march into Canaan."

She climbed the steps of the Memorial and took her assigned seat among a celebrated throng, some of whose faces she just recognized, others of those she knew well: Dr. King and Philip Randolph, Bayard Rustin, Roy Wilkins, Dr. Ralph Bunche of the United Nations, Thurgood Marshall, soon to be a U.S Supreme Court Justice; the young civil rights radicals Whitney Young, James Farmer, and John Lewis; Walter Reuther of the United Auto Workers Union; Rabbi Joachim Prinz; Bob Dylan, Odetta, Joan Baez, Marion Anderson—a formidable, endless list of famous names there to be counted.

Before the start of the proceedings, Dr. King leaned across to Mahalia and asked her to sing again his favorite spiritual, "I Been 'Buked and I Been Scorned," now one of Mahalia's trademarks. Perhaps, for King, that spiritual suggested the reason why so many poor blacks, at great personal sacrifice, had made the trip to Washington.

Mahalia recalled that she had begun the song in a traditional fashion, soft and gentle. Then something happened to her. "As I sang the words, I heard a great murmur come rolling back to me from the multitude below, and I sensed I had reached out and touched a chord. I was moved to shout for joy, and I did! I lifted the rhythm to a gospel beat. The great crowd began singing and clapping, and joy overflowed."

"When Dr. King arose, a great roar swelled from the crowd," reported the *New York Times* afterward. "He ignited the crowd with words that might have been written by the sad, brooding man enshrined within the Memorial."

Not only were the thousands in Washington on that sunny, invigorating day to hear for the first time Martin Luther King's sublimely beautiful speech of reconciliation, but the entire world would have his words to remember as one of the most important utterances of the twentieth century.

His speech built and bore the weight and passion of a symphony: "Even though we face the difficulties of today and tomorrow, I still have a dream . . . chiefly rooted in the American dream. I have a dream that one day this nation will rise up and live out the true meaning of its creed: 'We hold these truths to be self-evident, that all men are created equal.'

"I have a dream that one day on the red hills of Georgia, the sons of former slaves and the sons of former slave-owners will be able to sit together at the table of brother-hood.

"I have a dream that one day even the State of Mississippi, a state sweltering in the heat of injustice, sweltering with the heat of oppression, will be transformed into an oasis of freedom and justice.

"I have a dream that my four little children will one day live in a nation where they will not be judged by the color of their skin, but by the content of their character. . . .

"I have a dream that one day every valley shall be exalted, every hill and mountain shall be made low, the rough places will be made plain, and the crooked places will be made straight, and the glory of the Lord shall be revealed and all flesh shall see it together.

". . . From every state and every city, we will be able to speed up the day when all God's children, black men and white men, Jews and Gentiles, Protestants and Catholics, will be able to join hands and sing the words of that old Negro spiritual:

> Free at last! Free at last!
> Thank God Almighty,
> We are free at last!"

There is no doubt that through Mahalia's association with Dr. King and his movement, she saw the best part of herself playing a role in the changing of America. She had a definite place beside someone she looked upon as a living God. It enriched her and caused her to become a more politically sophisticated woman whose voice carried beyond the concert halls. She had acquired responsibility beyond her image as a gospel singer.

Years after Mahalia's death, her old friend and early supporter Studs Terkel recalled that mystical event, the second March on Washington, with deep affection: "Whoever was there can never forget that moment. Of course there was the address of Dr. King. His plea for sanity for mankind, for liberation for all. But preceding him, at his request, was Mahalia Jackson.

"I remember the great song she sang: 'I been 'buked and I been scorned / I been talked about, sure as you're born,' the glorious lyrics rising from one climax to another: 'I'm goin' to tell my Lord. . . .'

"By God, there was a great rolling murmur from the crowds! But at that moment, during the singing, there was an airplane buzzing. Don't ask me how it buzzed, or why, or for what reason. But it was buzzing around. And it was a noise that was quite disturbing. There were scores of thousands of people, remember. Hundreds of thousands around the reflecting pools. Behind Mahalia was the seated figure in stone of Abraham Lincoln. And there was that plane, and you might say, impiously buzzing. And she looked up. I knew what she was going to do. She just looked up at the plane, but she *sang* up to the plane. And, so help me, her voice drowned out the buzz! At that moment, the scores of thousands around the pool took out their white handkerchiefs and waved them as a wave of banners of triumph, for it was! She outsang, she, the human, had outsung the flying machine.

"And then the crowd cried: 'More, more!' and she could

not deny them. She stunned them with [the spiritual] 'How I Got Over,' driving herself and the audience into ecstasy:

> All night long, God sent his angels, watching
>   over me,
> And early this morning, God told his angels,
> God said, touch her in my name,
> And I rose this morning . . .

"That's just my memory of Mahalia," Terkel added. "I would say hers was a pretty full life. But I like Gwendolyn Brooks's poem—the poet laureate of our state, Illinois. Gwendolyn Brooks, the black poet, on the day Mahalia died, wrote "Mahalia Jackson: A Goodbye." In the mind of Gwendolyn, obviously, are the deaths of Martin Luther King, Jr., Malcolm X, Medgar Evers, and a friend of Mahalia's and mine, perhaps the greatest country blues singer of all time, Big Bill Broonzey.

"She wrote:

> 'Another break in the brick.
> No sorrow makes us immune.
> We lose and we lose again.
> The sorrows come and will come.
> Loss is forever.
> Again a goodbye.'"

Mahalia had summed up that extraordinary day in Washington: "We left in triumph, but we left as we had come—with peace and goodwill to all."

"How Long, Oh Lord?"

Mahalia was in her hotel room in Los Angeles, dressing for an afternoon rehearsal with Danny Kaye.

It was noon, November 22, 1963. The television set was on. The scene was a car caravan slowly making its way through the city of Dallas. There were President and Mrs. Kennedy—a scene, as it was every time they traveled together, of promise, power, and youthful magnetism.

In an instant the scene became a horror show. A nation watched as their President was gunned down by one or more sharpshooters.

Mahalia, as was true of the rest of us, sat transfixed before her hotel television set, in an obsessive, disbelieving stare as the repellant reruns of the murder were repeated and repeated, hoping somehow that the images were unreal, TV fiction. Praying that her adored President, for whom she had sung the "Star Spangled Banner" at his 1961 inauguration, could miraculously survive.

She was persuaded to appear that night on network television, to honor the nation's leader with the old Baptist hymn, "Nearer My God to Thee," her satiny dark features streaked with tears.

Her "Peace and goodwill to all" was short-lived.

Mahalia was fifty-one in 1963. During what can be described as a cruel travel schedule, her health began to be undermined, as Brother John had warned her might happen. She became a periodic resident of Chicago hospitals, suffering from exhaustion and a persistent heart ailment hardly helped by her lifestyle. She and Mildred Falls were large women who loved their rich black and Cajun cooking. They cooked big and ate big.

There were two major trips abroad planned, one to Europe especially focused on the Vatican, where she hoped to give a private concert for the Pope. It never took place, to her dismay. Then, with Mildred and a few friends, she flew to the Middle East—Egypt, Lebanon, Syria, and finally Israel, the Holy Land. She called the latter trip her second homecoming. There were final trips to India and Japan, and Europe again.

While at home in the U.S., Columbia Records, concerned about her health, scheduled multiple recording sessions, reasoning that she might soon be too ill to record anymore.

There had been many a night between trips back and forth to California when Mahalia was in a state of near-collapse. Her celebrated energy touched bottom. She reluctantly had to admit that she was bone-and-soul tired chasing fame and fortune. Brother John, always not far from her, despite their growing periods of hostilities, had known her too long, too well. Fame and fortune were the raison d'être in her character and had been from the time she was a teenager. He reminded her that she had been on a roll, hell-bent in her chase of the almighty dollar, and warned her that falling ill would be the only thing that could put the brakes on her drive. She admitted that she was insatiable.

"Girl," John said, "you got a long distance to go." But she knew more about herself than "Dear John." She knew she had overextended herself and didn't need his advice, that he was "cottoning" up to her, being solicitous, in one of their good times.

Mahalia was no reader, so she could not have read the seminal novel of Ralph Ellison, *Invisible Man*, published in 1952, a provocative charge against the estranged society in which black Americans lived their lives. And it is unlikely she read his public question, "Can any people live and develop for over three hundred years simply by *reacting?*" Ellison predicted the near future, when his fellow black citizens would face a storm-filled era in the sixties. Yet his prediction was to lag behind reality. Blacks in the northern industrial states were no longer the inarticulate minorities segregated in black diasporas. They were on the march, angry and daring, moving into the streets, into white enclaves, to places of employment, no longer simply reacting. They stampeded city halls to confront politicians head-on, and stirred up farragos of bitter trouble for previously uninvolved governments and private places of employment.

The response of whites to the demands of the blacks became bitter and bloody. Street-corner confrontations burst out like random firecrackers. There was a range of militant attitudes, a variety of choices for blacks for the first time: from Dr. King's non-violent resolutions to Malcolm X's aggressive nationalism and, ultimately, the birth of the Black Panther Party with its uncompromising program for political activism, encouraging their followers to take extreme and revolutionary actions, even if it were to mean facing police and National Guards head-on.

By 1965, Dr. King was focusing on the troubled northern tier of industrial cities—Chicago, Cleveland, Detroit—where the black migration from the Deep South had its terminus. If the horrors of black ghetto life were not to be eradicated, if the blighted lives of blacks were not to be altered, made more humane and fruitful, King knew his dream of a new America would be an illusion. He wrote: "Permissive crime in the ghettos is the nightmare of the slum family. Permissive crime is the name for the organized crime that flourishes in the ghetto—designed, directed and cultivated by the white national crime syndicates' operating numbers, narcotics and prostitution rackets freely in the protected sanctuaries of the ghetto. Because no one, including the police, cares particularly about ghetto crime, it pervades every area of life."

It was a serious political risk that Dr. King was taking with such public candor, and a popular black journalist, Louis Lomax, was to voice not only his own anxiety but that of many of King's colleagues: "By making a national public issue of the plight of Chicago's Negroes, Martin was on the verge of exposing not only a corrupt political system but the influence of the underworld in ghetto economic life as well. I was surprised that Martin did not disappear into Lake Michigan, his feet encased in concrete."

Three hundred years of white primacy was not about to retreat without battles and bloodshed, as it confronted rampaging black nationalism in Watts, California; Dr. King's failed open-housing campaign in Chicago; the sanitation workers' strike in Memphis; the street fights and fires in Detroit by the summer of 1967.

Mahalia received a phone call from Dr. King. It would seem that he was confronting more than white resistance in Chicago, and he needed her help. The local Chicago Baptist church leadership was putting up its own uncompromising resistance against Dr. King and the SCLC obtaining a church platform in the city. The malevolent opposition to King's presence had been orchestrated by the ultra-conservative Rev. Joseph H. Jackson, president of the National Baptist Convention. Jackson was no friend of King and his evolving political thesis that was moving beyond non-violent strategy. Doubtless, there was disagreement as to tactics in the post–civil rights movement, but the circumstances surrounding the Baptist church resistance suggested a jealous response to Dr. King's popularity, his dynamic leadership among whites now as well as blacks. There was little desire on Rev. Jackson's part to sublimate his authority to that of Dr. King's, and no wish to join the SCLC bandwagon as either colleague or deifier.

King knew that Mahalia had a strong if unaccountable alliance at the Chicago City Hall, with Mayor Richard Daley himself, who had yet to deny her any reasonable official request, particularly if it concerned his black constituents. From his own coming to power and her celebratory role in Chicago, Daley relied on her as an unofficial observer to report conditions of stress or any activity on the South Side that required the mayor's attention.

Following her conversation with King, Mahalia called Daley at once, and he agreed to come up with a suitable platform from which King could make a Chicago appearance, even if King was a man Daley disliked privately and was wary of publicly. Mahalia was able to assure King that Daley would provide the best platform in Chicago: the city-controlled auditorium at McCormick Place near the Loop.

The gala was set for the evening of May 27. King's plan, by now, was to override the South Side black church alliance that had distanced itself from him and his followers, and there were many who quickly came to his and the Gala's side: the SCLC Reverends Abernathy, Shuttleworth, and Walker from Birmingham, with Mahalia clucking like a mother hen to assemble an entertainment roster that included Dinah Washington (her young gospel rival), Eartha Kitt, the Chad Mitchell Trio, Dick Gregory, Aretha Franklin, and Al Hibbler. The Gala and the appearance of Dr. King was going to be a resounding slap at the Baptist preachers who had once welcomed Mahalia to their pulpits but were now slavishly following the leadership of a conservative, jealous minister.

Mayor Daley, the king and *the* political overlord of Chicago, was no liberal, no moralist, no enemy of the Mafia's special interests. He had said "yes" to Mahalia's request for King to be able to further his equal-housing-for-blacks campaign, but with King's invasion into Chicago's archaic political scene, Daley's private prejudices were stirred up. He not only resented but feared King's organizing of Chicago's blacks, the coalescing of their anger, their growing charges against *his* city. King was asking for an alliance between whites and blacks; he was making heart-breaking appeals for a new moral consciousness to the entire citizenry—an appeal few had heard in their lifetime. City Hall's response was silence. Young blacks rampaged through the streets. In the very depth of Chicago's classically hot and humid summer, Daley gave the implausible order to close all water hydrants in the ghettos. Young blacks were expected to get the Mayor's message: "Cool it!" Given the temperature of the air and soul of the ghettos, it was an arrogant and eventually destructive message. Four blacks were killed in street confrontations by an over-reacting Daley police force.

A Pandora's box had been pried open. Many followers of King and the SCLC, aroused by the reports of police brutality, decided reluctantly to take matters into their own hands—a break with King's temperate and non-violent philosophy. Daley declared himself against King's Freedom Movement and snidely told a group of Chicago reporters: "I'm not as non-violent a man as I used to be!" What he told Mahalia Jackson, as he moved against her friend, is not known.

The day that Daley made his headline remark—August 5—he ordered 1200 police into the black ghettos, allegedly to cool down tempers, but in reality it was to remind the people who was the boss of Chicago—Daley, not King.

Dr. King's strategies for non-violence were not working. Daley's power plays weren't working either. Blacks moved defiantly against the white communities that had abused them—there was to be no forgiveness. In nearby Cicero, white skinheads mobilized mobs to march under Nazi swastika banners of hate. Heads and bodies were cracked and broken.

It took Mayor Daley six months to conclude, in public, that *maybe* Chicagoans did deserve a policy of open housing. Until that moment of tentative truce, minorities had had no freedom of choice as to where they might live, non-segregated and within their means.

By 1968, President Lyndon Johnson's "Great Society" was coming apart. A federal commission led by the Illinois governor Otto Kerner made a serious study of the race rioting that was spreading across the nation. White racism, the study concluded, was at the heart of a national heritage of exclusion and discrimination, and the study warned that unless serious and radical changes were made by the government to guarantee justice to abused minorities, the nation would surely evolve into a separate, unequal, and undemocratic society.

When Johnson's failed "Great Society" was further corrupted by an intensified war in Vietnam, Andrew Young, then Atlanta's mayor and a colleague of Dr. King's, warned the President: "The bombs that you are dropping on Vietnam will explode at home in inflation and unemployment."

Martin Luther King, Jr., by now, had privately concluded that the civil rights movement might never reach fulfillment in the United States, given its established social system—it *was* and continued to be "separate, unequal, and undemocratic."

King's thinking had arrived at this point: Poor people's needs could never be met in a society where they were denied use of basic tools such as computers, and even if they were exposed to the techniques of capitalism, there was no room made for them in the workplace. It appeared to Dr. King that there was a designed deprivation of millions of people condemned to a level of existence of survival and no more. The poor, as minimum consumers, would contribute little more than their pain and humiliation, a failed family structure, and an economic disaster.

Given the extraordinary numbers of people in King's projected future of the nation, the inevitable solution, he believed, was to be premature death in the slums of America, or a revolutionary movement of the disenfranchised. Both King and the Kerner Report, in their separate quests for the terrible truth, called for answers that were not forthcoming.

King's evolving thoughts were making him a dangerous man.

The year 1968 became a double nightmare of political assassination.

The evening of April 3, Dr. King spoke from the pulpit of the Mason Temple in Memphis to an audience of followers. Eerily anticipating his own death, his voice filled with the sound now familiar to millions—a biblical, poetic, prophetic sound of exhortation and warning—he proclaimed that God had led him to the mountain: "I've looked over and I've seen the Promised Land! But I may not get there with you!"

The following twilight, while speaking with his staff on the balcony of the Lorraine motel before going to a soul-food supper preceding a mass meeting, Martin Luther King, Jr., was gunned down by a rifle shot aimed from a building across the street. He was thirty-nine years old.

A massive manhunt in the States, eastern Canada, and Great Britain led to the capture in England several weeks later of a James Earle Ray, a Tennessean and reputed narcotics and jewel smuggler. Ray confessed to the murder as he was being extradited to Memphis. He was tried and convicted for the assassination and received a life sentence. Ray acted alone, was the assertion of his attorney, Percy Forman, at Ray's sentencing.

Allegedly, James Earle Ray acted alone. His attorney, in support of this claim, asserted that both the Attorney General, Ramsey Clark, and FBI director J. Edgar Hoover had concluded almost at once that there was no evidence of a conspiracy. However, Ray's first attorney, Arthur Hanes, who had been dismissed by Ray, and the trial judge, believed that Ray did *not* act alone and that the Canadian police considered it "probable" that Ray had "important, perhaps underworld accomplices helping him make his escape through Canada." In 1978, the House Assassinations Committee asserted that "there is a likelihood that James Earle Ray assassinated Dr. King as a result of a conspiracy."

From the first days of the civil rights movement in the early 1950s, through the enlargement of the Vietnam War by President Johnson (of which King had been a decisive, public critic), notwithstanding King's accusations and challenges to the Mafia control of the black ghettos, it was indisputable that King did not want for enemies.

The day King was killed, Mahalia was in Nashville with a friend, Jean Childers, attending a meeting with Judge Benjamin Hooks, who was proposing going into the chicken-franchise business with Mahalia. During the meeting news of the murder was reported on television. Had Mahalia's visit with Judge Hooks taken less time, she would have been at the murder scene, as she had been scheduled to join King at an SCLC meeting in Memphis. In shock with grief and anger, Mahalia asked herself and friends: who and what was next in the United States? First Kennedy, then Malcolm, and now Martin! "America," she cried, "what have we become?"

She drove with friends from Nashville to Atlanta that day in a state of deep depression. They joined the funeral service at Dr. King's Ebenezer Baptist church. The entire directorship of the SCLC movement packed the chapel. Mahalia sat among the ministers. The rich, white liberal allies of King and the SCLC were conspicuously in view: Nelson Rockefeller, John Lindsay, and Robert Kennedy. As the service in Atlanta went its solemn, disquieting way, blacks all across the country were rioting. It was a climate of rising and militant expectation trading on the sadness and poignancy, the loss of a beloved leader.

King lay in state and the entire world paid attention, while the political reality of the victimized minority he had fought for passionately was finally being stripped of its illusions. Racial integration through non-violence—"love or perish"—now appeared as a chimera. More black pacifists than activists were to doubt seriously that King's politics of conciliation would survive him. Even before his coffin was lowered into the ground, newly arisen black groups were publicly condemning his appeals to reason; they were demanding a *quid pro quo* for his murder.

Some supporters now recalled the profoundly moral caveat he released while imprisoned in the Birmingham jail five years before. He had dared to ask the question in his famous "Letter to the World": "Who was the Negro's greatest enemy?"

"Not the White Citizen's Councillor, or the Ku Klux Klanner, but the white moderate who is more devoted to 'order' than to justice, who prefers a 'negative' peace which is the absence of tension, to a 'positive' peace which is the presence of justice!"

America's political bloodbath resumed on June 4, 1968. At midnight, while celebrating his primary victories across the country, and addressing a crowd of campaign workers in a ballroom of the Ambassador Hotel in Los Angeles, Senator Robert Kennedy, candidate for the presidency, became the next target of an assassin's bullet. An unknown man, as anonymous as James Earle Ray, a man named Sirhan Sirhan, pointed a gun at Kennedy and fired point-blank. Kennedy was mortally wounded and five bystanders were hit. Yet another blood-curdling, stomach-turning, mind-blowing video image was to be indelibly impressed on the reeling psyche of the nation. A candidate who had been gaining the trust of the poor, the minorities, the civil rights advocates, had been sacrificed again. Robert Kennedy had moved out of the shadow of his brother's presidency, become his own man; he had gained maturity and insight into the crucial politics of necessary social justice by the time he ran for the presidency himself. One could say that he had become another dangerous man for those who were opposed to fundamental economic and social change.

Subsequent investigations have suggested that a second assassin was present in the ballroom, a man who got close enough to Kennedy, with what observers believe was a gun in hand. Kennedy, on being hit and while falling to the floor, had grappled with him, pulling off his clip-on tie from around his neck. In the confusion, the second man had disappeared into the panicked crowd.

Now, unhinged by the quartet of murders of national figures—the two Kennedys, Malcolm X, and King—many in the country feared that an organized conspiracy existed. In that shocking time for the democracy, it was distressing to think that the powerful police agencies of the federal government had failed to uncover leads to *any* covert conspiracies beyond the alleged lone assassins.

Mahalia Jackson had met and admired the young ex–Attorney General during the Washington Rally with Dr. King. Filled with sadness and the panic felt by the whole nation, Mahalia pulled herself together the next evening to leave her hotel room and play a part in the CBS television memorial program for Robert Kennedy. With tears cascading down her cheeks, her voice a cry of misery, she sang "I Been 'Buked and I Been Scorned" *a capella* (the only gospel she ever sang without accompaniment, as she had done for Dr. King at the Washington March), and offered up an act of devotion. She did not attend the mass for Kennedy held in New York's St. Patrick's Cathedral, but remained in her hotel room and prayed for him in private.

"*Didn't It Rain?*"

After her failed marriage to Ike Hockenhull in the early forties, Mahalia appeared to put serious relations with men out of
her mind. But when she would return home at night after an
exhausting concert, she never appreciated the empty rooms,
and knew she would have to be in search of another voice in
her life. Whenever she sang in churches or concerts, largely
attended by black audiences, she would often make a flam-
boyant entrance flouncing her long skirts, look out at the
audience and, with a giggle, say: "Out of all the good-looking
men I see here tonight, I ought to be able to find myself a
husband!"

She told friends that she made the comment as a joke, but her intimates knew otherwise. She had a healthy appreciation of the male sex, needed companionship, and confided to women friends that she was indeed in the market again.

In the early forties when she sang in a church in Gary, Indiana, the Galloway family had befriended her. The husband, Sigmund, nicknamed "Minters," she had found exceedingly attractive, but he had a wife and daughter. In the next few years, his wife died suddenly, and his daughter went to live with an aunt. Sigmund had made his living working for an uncle in Gary, who was a house builder. When his wife died, Minters took off for Southern California to take up his true love—music, both popular and church. He formed a five-piece "combo" in which he played the sax and flute and was successful for a decade in the Los Angeles area. As Mahalia's personal appearances took her to the Coast a number of times a year, she and Minters reunited, and he soon became her exclusive escort at Hollywood dinners and parties. In 1964, when she returned from a trip abroad, he moved to Chicago, and from that time they were a courting couple. Before the year was out, they were married in Mahalia's living room by the pastor of the Greater Salem Baptist Church, the Reverend Leon Jenkins. Two friends were their witnesses, joining them at a small wedding dinner in the Loop.

Minters served Mahalia's career briefly. He played his flute at numerous recording dates, but the mood of the marriage from the start was contentious, a constant barrage of antagonism. Minters saw himself becoming her manager and director, taking charge of her career, all of which she found intrusive and unacceptable.

He said she nagged too much.

She said he liked Chivas Regal and other women too much.

He said she was too controlling, a commanding woman no man could please.

She said he tried to hit her and broke his hand in the attempt.

And she didn't care one bit for the cruel look in his eyes when he turned on her.

Finally, fed up with his demands for cash and forgiveness, Mahalia moved out of her house into the Chicago Sheraton Hotel, then to a small place on the Lake Shore. Soon after, she asked her lawyers to draw up divorce papers.

During the negotiating period of the lawyers, Minters refused to leave her home and ran up huge bills. Mahalia wanted out, no matter the price. In 1967, the divorce became final.

In the year of the divorce, Mahalia had received and accepted an invitation to perform at New York's Lincoln Center, in a concert called "Salute to Black Women." She could have stayed in a hotel, and many friends extended invitations to put her up, but she chose to stay with Brother John Sellers, who had a three-room apartment on 65th Street and West End Avenue, a stone's throw away from Lincoln Center. When she moved in for her New York stay, she told John straightout that she wanted to pay him $150 a week. She also announced that she had invited Professor Theodore Frye, a Chicago choir director identified with her early career, particularly her song "Move On Up a Little Higher," and her cousin Alan Clarke, who sometimes acted as her social secretary, to join her.

John stared at her. She had gone out of her head. "How do you figure all of us are goin' to live here? I've got just three rooms. Mahalia, you're crazy!"

She was unruffled by his temper. She told him she had

decided to buy a cot. "Someone, I guess, can sleep on the cot," she said.

John's retort was: "Mahalia, we all can't stay here. I'm not used to things like this. I don't want those men here!"

"Well," she replied. "Oh, Nigger, don't you be so grand! That's what's the matter with you, you're too grand!"

She came to New York with $40,000 in cash. When she arrived in John's apartment she removed the bills from her oversized pocketbook and counted them out, placing the piles in John's dresser drawer. He recalls with a shrug and a roll of his eyes that she was then making $1000 to $1500 a night when she sang in churches.

Whatever Mahalia wants, Mahalia gets. She, John, the Professor, and the cousin somehow managed to fit themselves into John's apartment. When the Lincoln Center concert was over, her first act of packing was to check her money in the dresser drawer.

As John tells it: "Slyly, she looked at me while she counted the bills, saying, 'I know you wouldn't take anything from me, but I'm goin' to count it anyway.'

"'Yeah,' I said. 'Count it, because I want only what belongs to me, and when you've done it, you pay me the money you owe me for staying here for three weeks.'"

Mahalia looked up at him. "'Boy, I'm missing one hundred dollars!'"

John said he laughed and answered, "'You can't be missing it unless your cousin stole it from you. Alan Clarke is nothin' but a thief! Count it again, count it over, and let me help you count it. Here's what you owe *me*—four hundred and fifty dollars for the three weeks of everybody livin' here.'"

Mahalia objected, continuing their game for a lifetime of a mismatched, embattled surrogate mother and son, saying she hadn't earned much money and couldn't afford to pay what he asked, although she had offered it when she arrived.

"'Shit, don't tell me what you ain't made!'" John had responded, with a deliberate vernacular fury. "'I've been sittin' up there and seen what you did at Lincoln Center! Look at the mess of gospel books you sold, and almost two thousand dollars' worth of sheet music!'" She had gotten the books from Thomas A. Dorsey and would split the profits with him, as she had done for years. John went on to explain the finances of the traveling gospel singers—how they made their expenses in the early days selling sheet music—yet Mahalia, who had by now sung for kings and queens, was "still doin' it."

Finally, Mahalia agreed that the entire $40,000 was intact. Why she played her little game with John, only she and the Lord knew. Now it came time to pay John the money she had promised him for putting her up, as well as her two "guests." She started to count out the bills, saying: "'I told you I was going to give you a hundred a week.'"

John exploded. "'Oh no, darlin', you said you'd pay me one hundred fifty each week. You *know* I hear well. Let's not fall out over money, Mahalia. We've been friends too long to fall out over a can of pork an' beans. Your cousin Clarke told you a damn lie! He said that I said you had eaten everything in the house, and didn't want to put anything back in the house or buy nothin'. Now we fall out over that—a twenty-five-cent can of pork and beans! And you got all this money! You know, you all ought not to be here no way. You all ought to have gone to a hotel!'"

It was Mahalia's turn to get mad: "'I *will* go to a hotel the next time!'"

John: "'In the first place, it was only supposed to be you.'"

Mahalia: "'Well, three hundred dollars is all I can spare.'"

John sighed. "'All right, give me the three hundred, but let me tell you this: I'll never do nothin' with you again as long as I live.'"

She exploded. "'Nigger! I got money enough for you to carry out my shit for the rest of your days!'"

"'Black bitch! You will never live that long. You've got the nerve to tell me something like that in *my* house, after I've spent my own money feeding you. You've got to be crazy, or you've become nigger-rich! But you'll never live to spend it.'"

Mahalia got to her feet, looked around for her pocketbook, and then answered him. "'Well, I'm going now.'" She busied herself packing her things, not so much as giving him a glance.

John stood by, watching her. "'Mahalia, you're cheap,'" he remembers saying, recalling the end of that painful memory scene. "'It was always that way. If I got a bit more applause than you, you didn't like it. You know, you're a jealous bitch!'"

They didn't look at each other after that. She moved around him with her luggage, with some difficulty. He made no move to help her, opened his door and let her out. Then he slammed it closed behind her, swearing to himself that he would never speak to her again.

"It's Me,
Oh Lord"

Benjamin L. Hooks, the personification of the NAACP since the 1970s, was a young law student at DePaul University in Chicago when he made the decision of his life. There were serious rumblings that civil strife was building in the southern states, and Hooks, after going home to Memphis and conferring with family and friends—the year was 1948—left law school and joined the cause in that city.

He deliberately flung himself into the battle, offering his services as a junior legal adviser to blacks in police custody, and in incidents of a discriminatory circumstance.

In the Memphis courtrooms, Hooks was offended by the

practice of disparaging black lawyers by never referring to or addressing them as "lawyer" or "attorney." In addition, special benches were set aside for black attorneys, reminiscent of the early Hitler years in Germany, when certain park benches were marked with the Star of David and *reserved* for the pariah Jew.

Hooks moved into the civil rights movement quickly as a lawyer and businessman. In the sixties, he became a Tennessee state judge appointed by the liberal governor Frank C. Clement. In those years, Hooks also served on Sundays as a Baptist preacher in Memphis, and occasionally in Detroit. It was in his capacity as preacher that he came to know Mahalia Jackson, becoming her fan and partisan. He recognized her native intelligence, her shrewd instincts, and mother wit.

Mahalia, he said, "was a woman who'd been battered and mistreated, someone who'd known negative treatment women in America had come to expect. But in her case she had the added, and dubious, distinction of being a black woman!

"She had been hoodwinked, defrauded, and beaten out of money. She'd give concerts, you know, and at the close of it not get paid her money. She got to the place where she'd refuse to sing the second half of the program until she'd gotten the money due her *in hand,* in cash or cashier's checks. I've been with her on those tours," Hooks reminisced, "when the house was full, and the promoter would come up to her and say: 'Now, we'll just send you our check.' And Mahalia replied: 'Oh no! I'm not goin' back out there until either I have my money in cash or a cashier's check, mister.' But, you know, at night in those small towns, it's pretty hard for anyone to get a cashier's check anywhere.

"And I remember her on so many nights when she'd take that big roll of money from the promoter and stuff it right down into her bra. And then she'd sweep out onto the stage and sing like a mockingbird! Yes, she wasn't ever going to get hoodwinked ever again."

Hooks was impressed by Mahalia's shrewd and suspicious nature, her form and style in the act of survival. She had come to believe that being suspicious of people's motives was normal and natural. Even her politics, Hooks commented, were all related to her life and art in the Baptist church. That's where the greatest politicians in the country were, he said cynically. Preachers, deacons, and lay people.

"The white people kept us out of secular politics," Hooks mused, as we sat for five hours in the belly of a hotel in Manhattan, recalling a time and his role in it. "We were, by force of nature, compelled to exercise whatever gifts we had in our lodges and churches. There was an inordinate amount of politics in the black church that has now subsided because we now have other outlets. But when you've been mistreated all the week, called "boy," given inadequate wages, Sunday morning was the only time you were a man. (He was speaking of himself, as well.) Manhood and womanhood exerted itself! and it was no surprise to me that Martin King—son of a Baptist preacher—Ralph Abernathy, Jesse Jackson, Andy Young and myself, we were very familiar with politics because you just practiced them on a different stage. To me, politics was how you compromised without yielding your principles.

"Now let me tell you how I came to know Mahalia. There were a couple of white boys in Nashville by the name of John and Henry Hooker. John had run for the governorship of the state of Tennessee. The Hookers and other whites from Fisk University started an organization to get black people into the voting process. In Memphis we voted through the years, but we were more *voted* than *voting!*"

Black people vigorously wanted this white man to be governor, and 90 percent voted for John Hooker, but he lost anyway. Ben Hooks made the comment that the Hooker brothers were the most unbiased white men he ever met. When John failed to make the governorship, he and his brother Henry decided to go into the chicken-franchise business, which they named "The Minnie Pearl Chicken System" after the white southern country singer. They offered Judge Ben Hooks and his partner, A. W. Willis, an opportunity to join them in the franchise business, even to help finance their start-up costs.

Hooks and Willis decided they wanted their own business rather than to depend on a franchise from the "Minnie Pearl" outfit.

Watts said to Hooks: "'Ben, you're a preacher. Why don't you talk to Mahalia Jackson? She's a more important theatrical name than Minnie Pearl. If Mahalia agrees to work with us, then we can start our own—The Mahalia Jackson Chicken System.'"

In those years (1967–68), there were no franchise companies in the South that entertained the idea of doing business with a black person. But the Hooker brothers agreed to work with Ben Hooks and Watts, particularly if Mahalia accepted their plan.

By now, Mahalia was the best-known, most luminous name in black society. Hooks went to see her in Chicago, and because he was both a preacher and judge, they got on very well.

When Ben Hooks met Mahalia, her second marriage had come to an end. She was still seeing her first husband, Ike. She had also just made an extravagant purchase of two condominiums on the South Side of Chicago, turning them into one huge apartment. Hooks was impressed, but it was her business acumen that impressed him most. "They were the sort of questions you'd expect from a Wall Street lawyer!

"Mahalia agreed to become a partner, insisting that she have a 15 percent share of ownership in 'the business' and be paid a fee of $10,000 each time we added a new store to the company."

Six months later the firm opened its first store in Memphis, and Mahalia made a triumphant appearance. "You couldn't contain the crowds," Hooks recalled. "Even when she didn't sing, just her presence was enough. The crowds packed the place like they would today for a Michael Jackson.

"Hooker opened nearly 400 stores of the 'Minnie Pearl' system. He then bought a chain of stores in Florida. Our 'Mahalia Jackson' subsidiary numbered 135 stores eventually. The Hookers decided to go public. On paper, they were worth over one hundred million dollars!

"Our system was building faster than the profits from the internal operations would allow it to grow. We were going public too, but didn't, being persuaded by our broker that it wasn't the right time."

Soon Hooks and Watts added another company to the "Mahalia Jackson" subsidiary; the "Mahalia Jackson Food System," and for it they produced some twenty-five varieties of food—peas, beans, and corn, for canning. They worked with the A&P chain, which in those years owned and operated a great number of canning plants with their own name, Ann Page. Hooks made an agreement with A&P to switch labels of millions of cans of vegetables and fruits on the production line. It took but a few minutes to change the label from Ann Page to "Mahalia Jackson System," complete with a picture of Mahalia.

Dr. Hooks noted that the franchise operations of both companies came under some criticism from the U.S. Securities and Exchange Commission some years later, because of their accounting procedures, and put the reins on both systems. The companies eventually went bankrupt about a year following Mahalia's death.

Ben Hooks felt deeply that Dr. King and his political and moral impact on the national scene was the turning point for Mahalia in her last years. "She loved Martin—and I mean it, of course, in the good spiritual sense. The ground he walked on, to her, was holy. Martin King represented what she wanted to be, and what she would have given her life to accomplish."

He recounted an incident that took place between Mahalia and King at the famous Washington March. It can't be confirmed, but it carries the cachet of truth, given the two personalities and how passionately they valued each other's gifts: "A short time before that March, Martin had been in Detroit with Rev. C. L. Franklin, Aretha's father. Franklin had led a big march through the city seen by hundreds of thousands of people. It was there, in Detroit, that Martin first delivered his immortal 'I have a dream' speech.

"The story goes that later, at the Washington Monument, when Martin was prepared to give a written speech previously approved by the Committee, Mahalia, sitting near him, leaned over and whispered, 'Martin, give 'em the *I have a dream*. And she may have pushed him on in the Baptist tradition, to deliver, if not the Detroit speech, at least the ending of it. Well, he did it."

As for Mahalia's lifelong identification with the old-time Sanctified Baptist church service, Hooks said: "The style, you know, is to hold onto the notes for a long time, and change the voice to a falsetto. It's been done before and copied ever since, but no one ever did it like Mahalia. Her favorite preachers were always the ones who preached with fire, power and spirit. Now, here was a woman who was not poverty-stricken. Mahalia left a good estate, but there is no telling how much money she could have made had she gone into the secular world. Most of the early black artists were raised in the church and moved over. Mahalia never lowered her standards in terms of what she believed. Her veneration of the ministry transcended almost any human expectation. I often wondered, even now, what would have happened if she'd had formal voice training. It might have ruined her!

"Since her death I've listened to all the people who call themselves the kings and queens of gospel music, but they can't touch her—side, top, bottom—they just can't do it!"

By 1967 Mahalia had moved from her Prairie Avenue house to the double condominium with eleven rooms on Cornell Avenue, near the Lake.

She and Brother John had become reconciled since their falling-out in his New York apartment. Their lifelong tempest was able to survive every psychological and financial onslaught, it seems, for they made believe there had never been the ugly words between them. The memory of the New York derailment had faded away. John was now being paid $500 a week by Mahalia to help fix up the huge residence that was to become a functional business office as well for her increasing professional interests. He was also supposed to cook (he was a great chef as well as singer, then temporarily out of work; she knew very well what having him around would mean— delicious fare). John accepted the job offer, knowing or hoping it would only be for six months or less; it obviously pleased him to be of help to her, be part of her life and close to her gospel art and business, no matter what the personal cost; it was their fate to touch, to go, and to come together again.

Mildred Falls, by now, had been cut off from Mahalia's circle and was rarely sought out or seen by any of the old crowd, except for Brother John. He decided to telephone her and gossip about his and Mahalia's reconciliation, which Mildred found almost unbelievable. She laughed, "Well, you know as well as anybody how Mahalia is when it comes to money." John told her that Mahalia had agreed to pay him $500 a week while he worked for her in the condo. "But," he added, his laughter joining hers, "I sure hope I don't have to throw her out of the 31st floor window here, if she messes with me about money again!"

Now, Mahalia was about to shock her lifelong black law firm in the Chicago Loop by firing them. She informed them by phone that she had decided to transfer all of her business affairs to a young Chicago lawyer, Eugene Shapiro, whom she had only recently met. She ordered the Estridge firm to deliver her files and any new contracts to the "new man" immediately. Her old firm was dumbfounded, for the business of overseeing her recordings and concerts at home and abroad had been taking up a good portion of the firm's time, with the volume of her legal work now ranging in the hundreds of thousands of dollars a year. One of the shocked law partners called her back to ask her if she had gone off her rocker. "Did we hear you right, Mahalia? Are you telling your old lawyers and friends we're through, you and us?"

"Well, Mr. Ming," she said, "that's the way I want it. Turn over my business, the files and the bank business, to this man. And that's that."

Her move—from a recognized firm of black lawyers who had managed her legal matters all of her professional life to that of a young white counselor—shocked her family and friends. Her behavior made no sense to them, but she was adamant.

How to explain what was probably an irrational act? To abandon one's friends and advisers not when one was failing, but at the zenith of success? She had become acquainted with a woman whose husband, a black architect in Chicago, used the legal services of the young lawyer. Mahalia was under great stress from her fast-paced schedule when suddenly the wife of the architect became her constant companion, ingratiating, flattering, and advising, according to Mahalia's astonished old crowd. Some have even speculated that the architect's wife had offered to help the young lawyer turn Mahalia in his direction—with the large fees and the notoriety of her stardom following.

Once the transfer of lawyers took place, Shapiro (nicknamed "Superior" by his client and her friends) and a small staff suddenly appeared and set up their office in Mahalia's condo, beginning work at ten in the morning while Mahalia was still abed, according to Brother John, himself a member of her "staff." He recalls walking into their office area, with its piles of contracts, mail, and promotion photos on the desks, mumbling a "good morning" and heading off for the kitchen, wondering whether the new lawyers were paying their sleeping client for all the space she'd given them. Knowing her so well, it probably made her feel important to be able to look into the "office" and see her very own legal staff hard at work on her business, smack in her own home— ultimately, the workings of a whole empire under her very nose.

That was also the year in which she began to fight depression and ill-health stirred-up by the stress of her schedule, the year she went in search of the skin lighteners that Chicago salons were offering their black customers.

"No, it was not a very good year," Brother John grimaced. "And they sure messed up her face! No, it was not a good year, and I watched it. I was a fly on the wall that year."

From the time that Mahalia joined the roster of Columbia Records as "the world's greatest gospel singer"—her giant step into the white world—it was asked of her that she Anglify the traditional orchestrations and arrangements of gospel songs, the blood-and-bone of the black church that had emerged from Reconstruction. She resisted this white managerial bid to lighten the cries and passions of race, but, like most performers, she had to humble and cool her art occasionally and succumb to the ever-present commercial pressure, conceding more than she ever wished.

To the credit of her art, Mahalia sang popular songs when urged to do so, many for recordings and telecasts. To each, she provided the ultimate in the gospel sound of unforgettable spirituality: songs like "Danny Boy," "Summertime," "Sunrise, Sunset," "Trees," "Lonesome Road," "Without a Song," and "The House I Live In."

Through the years, she paid an ever-accumulating debt to achieve stardom. Her dual loyalties: first, to her rapacious self-aggrandizement; then, to the music of her people that was reflected in her distrust of the society that had scorned and rebuked her race for three hundred years, as well as her pitiless drive to become part of the growing rebellion aligned against it. Such cruel, internalized injuries can never be computed.

Becoming a preacher in her own temple was Mahalia's long-held dream. From her arrival on the South Side of Chicago, her idol had been a preacher, Elder Lucy Smith, a clever woman who solicited contributions for the building of her own church from 1934 on. It was well known that her parishioners were not only induced to contribute into the building fund but actually helped build the edifice. Elder Lucy couldn't read or write, but she had an astonishing gift of persuasion, able to induce the faithful and the unfaithful who entered her sphere, including many whites, to give her gifts of money and bundles of old clothes for "the deserving poor." Elder Lucy was the first woman preacher in the Chicago area who spoke to her congregation and the community by means of radio. While Elder Lucy ran the church with her granddaughter, known as Little Lucy, her daughter, Idella Smith, did all the radio announcing because she was "educated."

Elder Lucy was noted for her personal appearances, driving slowly about the South Side of Chicago in a huge, low-slung limousine, nodding this way and that to the passersby. She was also famous for her proclamation that, when she died, she wanted to be carried to the graveyard in a carriage drawn by two big, white horses, because—she'd say—she didn't want to smell gasoline while being lowered into her grave. Mahalia adored the old lady, both for her dynamic presumptions and for her recognized public works for the ghetto poor. As a young woman, Mahalia also envied the old lady her church and notoriety. "If I ever get me some big money, I'm going to have me a temple!" she proclaimed. And true to her word, when she began to accumulate sizable capital from the royalties of "Move On Up a Little Higher" and her personal appearances, Mahalia overcame her shyness in approaching people for whatever she wanted, whites as easily as blacks, professional associates or strangers; nothing fazed her when it came to her ultimate ambition: to have her own temple, a living monument to mark her earthly achievements.

But she didn't live long enough to see and taste the ultimate fruits of her hard work. She had waited too long. People from all walks of life were frequently generous when she spoke of her dream, giving her cash and checks for the projected temple, asking nothing in return, for Mahalia had become a super-salesperson. There was even a story (part of it authenticated) that Nelson Rockefeller, when governor of New York, had become a Mahalia Jackson addict after watching her on television and in concerts and played her LPs in the governor's mansion repeatedly. When she performed at the Brooklyn Academy of Music, he attended the concert with bodyguards and finally met her in person—a meeting that was recorded by press photographers. The rest of the story could be apocryphal, but characteristic and fondly spoofing her money attitude: that Mahalia told Rockefeller of her temple plans and some time later he sent his brother David to deliver a $10,000 check for the venture. It was reported that she had expressed dismay that it wasn't in cash, but accepted whatever cash he had, and his promise that the rest would also be delivered to her in currency, if that was her wish. She then was said to have stuffed the first contribution (all the money David Rockefeller happened to have on him that day) into her handbag. She thanked him kindly. Glancing at the friend who was supposed to have witnessed the transaction, she gave a look that said: See how I get the big shots of this world to help me build my temple!

She had obviously been convincing when she proclaimed her intentions of building the temple, a monument to her brothers and sisters, those who had come before and those who were yet to come, she would explain, who, like herself, had journeyed to the Promised Land.

Dirge
for
Mahalia

On January 27, 1972, Mahalia Jackson died of a heart seizure. Only sixty, she had burned her heart out. Medically unstable for years, and often taken ill on stage, during a performance, it is probably safe to say that her glorious vibrato had succumbed before she relinquished her body and soul to her God, in the Little Company of Mary Hospital in Evergreen Park, a Chicago suburb.

Her death was to mark the eclipse of gospel's golden age, not produced by her alone—there were other popular and impressive talents of the art who survived her—but Mahalia seemed to possess a magic elixir which Nature had denied her competitors.

Two weeks before her death she sat up in her hospital bed and told her secretary, Celestine Fletcher, to get a pencil and write down on the back of an old envelope a shorthand message for all her well-wishers. The code read: "Jan. 11, 1972. 6:35 pm. Psalm 119. Verses 17 and 18."

Later, when she was alone, Mrs. Fletcher looked up the reference in the Old Testament, and it read: "Deal bountifully with Thy servant, that I may live and keep Thy word. Open Thou mine eyes, that I may behold wondrous things out of the law."

After Mahalia's death, Mrs. Fletcher said she had been consoled. "You go read those verses," she advised family and friends. "They're haunting me still. I guess Mahalia knew something all the rest of us didn't know."

While Mahalia lay in state in her coffin, in the Greater Salem Baptist Church where she had given the gifts of her art and soul for forty years, life in Chicago came to a virtual standstill.

Thousands of mourners assembled behind police barricades the night before the funeral, milling about to keep warm on the snow-covered streets of the South Side. Later, many could never figure out how they had survived the night, but there were few who wanted to risk losing their place in line, when the church doors were opened in the morning. Those who had endured pressed against one another as they were swept into the building, down the aisle to the altar on which the rose-covered coffin sat imperiously.

Mahalia's large, expressive face was fixed in a smile. She had been dressed in a long blue gown with golden threads; her black beehive was piled high on her head. Her hands, encased in white gloves, held a large Bible opened to the 20th Psalm of

David: "The Lord hear thee in the day of trouble; the name of the God of Jacob defend thee. . . ."

Some mourners, as they passed the bier, whispered private prayers; others, their faces streaked with tears, mouthed fragments of gospel songs. They all stared down at the tableau on the altar with expressions of disbelief—the queen was dead, a portion of their sacred lives had gone away, left them.

The organist, LeRoy Dulley, was transported in a private grief. He stood up and shouted: "She's found!"

The crowd shouted back: "She's found!"

Toward the close of the afternoon, the crowds thinned, and Mayor Richard J. Daley entered the church, accompanied by three bodyguards. He walked royally toward the coffin, leaned over as if to speak to Mahalia, then turned to acknowledge his constituents, whispered remarks to the ministers standing respectfully about the coffin, and made his way out, his duty accomplished.

The church filled up once more, and the Jackson family made its entrance, as if to trumpets, with varying degrees of stoicism and tears—brothers, sisters, cousins, and Aunt Hannah, the great survivor. To one side, uncomfortably, stood Mahalia's last and current attorney, Eugene Shapiro, and, close by, her previous and now rejected attorneys, Robert Ming, Aldous Mitchell, and Chauncey Eskridge. And former husbands Sigmund Galloway and wheelchair-ridden Ike Hockenhull were there. Also, somewhere in the crowd was Brother John, thinking about what he would sing as his farewell to her the next morning.

When the Jackson family arrived, the congregation singing and humming gospels halted on a dime, as if to show respect to the most important mourners. But Aunt Hannah was heard to cry out in a magisterial voice—"No tears! No tears!"—the congregation took it as a signal to resume their singing.

The Chicago press and radio stations announced that Mayor Daley was extending the memorial service for another day. The large Arie Crown Theatre on McCormick Place, Lake Shore Drive, would receive the coffin for display, for tens of thousands of people had complained that the church was too small to handle those who wanted to pay their respects and pray for Mahalia.

Even before the Arie Crown service got under way the next day, the Reverend Jesse Jackson conducted a radio memorial service, invoking her name and praying: "I give thanks to God for letting Mahalia come this way." He was followed by Alberta Walker singing Mahalia's emblematic song, "Move On Up a Little Higher."

The Arie Crown Theatre memorial, drawing a crowd of 10,000 mourners, began with Coretta King being given a standing ovation. She gave thanks to Mahalia for being "black and proud and beautiful. . . . A woman with extraordinary gifts as a singer, singing songs of her people. She was my friend and she was the friend of mankind." Coretta's eyes filled with tears. She recalled how her husband had often said of "this lady, 'A voice like this one comes not once a century, but once a millenium.'"

"Death is a mean man!" the Reverend Joseph H. Jackson said, as he gave the principal eulogy. "I pray that my fellow mourners and the grass and the breezes be gentle with my friend." He then said in a loud whisper, so that the whole congregation of more than 6000 could hear him: "Let her alone!" Thus, the once adversary of Dr. King when Mahalia brought him to Chicago.

A sea of voices answered: "Amen."

Then, Ella Fitzgerald: "This great woman is now gone

from us. I tell you, there's not another singer like her in the world!"

Brother John Sellers, powerful gospel singer in his own right, as well as Mahalia's calling him her godson from the time he was a boy, could never have imagined his being excluded from the service, denied the opportunity to sing and affirm his bond and admiration. Twenty years later, he still smarts from the scenario designed by Mahalia's secretary, Celestine Fletcher, who took over the details of the uptown service, filling a vacuum no one else appeared ready to assume. She had replaced him with Aretha Franklin. It was not yet sunup when John received a shocked and angry telephone call from Studs Terkel, who had only moments before been told of the switch in the program. He threatened to walk away from the service himself, for he felt the switch to be an aberration.

Gossip enters, and gossip often is the truth. Mahalia and Aretha had been friends when Aretha was a rising gospel star who looked to the older woman for direction. But that was long before Aretha's father, the celebrated Baptist minister in Detroit, got himself into a mess of trouble and off Mahalia's list of desirables.

It would appear that the Reverend C. L. Franklin was a popular religious leader in the Detroit-Cleveland-Chicago axis, when he began to make numerous and mysterious trips to northern Colombia, South America. His trips were particularly mysterious to the FBI and Customs, for there was little evidence that the trips were necessary or personal. In his own parish, he had the reputation of being a ladies' man, so in time his alleged drugs-and-sex activities became local scandal, whispered wonderment in his congregation. By the time the government decided to become curious about his peregrinations abroad, a gang of local drug dealers decided they wanted to share in whatever business he was doing; they fired a warning shot that he took seriously, but he evidently made no move to do business with them. The government got around to indicting him for smuggling, and, in search of character witnesses for her father, Aretha went to Mahalia, asked her to appear at the trial for him. Mahalia, it is said, refused, saying she knew the Reverend "snuffed the white stuff, and besides, everybody knew he was a womanizer."

Aretha felt abused by Mahalia's attack and retaliated with: "You don't have to be so grand with me, bitch!" From that day, to the day of the funeral, it is believed that they never spoke.

Yet, here was Aretha! And here also was her father. While Brother John Sellers had been banished! It was bizarre.

Terkel deemed it bizarre, too, and thought long and hard about making a Chicago scandal, but didn't (and spoke briefly). And neither did John, though the hurt was bitter and painful.

There is a conclusion to the Reverend Franklin story: A few years after Mahalia's funeral, the man was again pursued by the Detroit gang that insisted on a cut in his alleged drug business. One night they succeeded in gaining entrance to his apartment, lay in wait for him, and when he came home he was met with a hail of bullets to the head, leaving him not dead but permanently brain-damaged for seven years before his own death.

In her turn in the funeral service for Mahalia, at the Arie Crown Theatre, Aretha sang "Precious Lord, Take My Hand," composed by everyone's mentor, Thomas A. Dorsey. She worked the hymn higher and higher, with her voice bounding off the theater walls, as if to prevail against all the evil rumors. Her father, not to be outdone by his daughter, took his turn at the podium: "She has moved to her other house," he shouted. And thousands of mourners answered with amens. "We came here to rejoice," he rejoindered. "She laid down her heavy burden."

The funeral service at the Arie Crown Theatre ended; the coffin lid was closed. A news photographer approached Clara Ward, the gospel singer, and asked if she would smile for her picture next to the rose-covered coffin. She glanced over at the box and shook her head. "No, Mister. This is not a good day for me to be smiling."

By late afternoon the coffin was transported, along with a garden of flowers, to O'Hare Airport, and it was flown to New Orleans. Mahalia had come home.

Distressed and crying people jammed the airfield in New Orleans. The cortège moved slowly from the airport into the city, through the black "inner city," along the railroad tracks, in sight of Old Mississippi. The sidewalks resounded with mourning tears, amens, and cries. Their native daughter had come home again. Their queen, she was dead. Discreetly leading the way to the giant Rivergate convention hall was a squad of New Orleans police motorcycles, and inside the Center were five thousand New Orleans citizens gathered to pay their last respects to one of theirs. They chanted and clapped hands in time with the five-hundred-member city-wide choir. With the Jackson family sitting down front, close to the rose-covered mahogany coffin, they led the immense crowd in thunderous hand-clapping, foot-stomping, as Betsy Griffin, a local gospel performer, sang "Move On Up." A color guard of three marines and a man from each of the services flanked the coffin. Ministers outdid one another as they glorified and adored their blessed Mahalia, born between Water and Audubon streets, "back'a town" sixty years ago.

Outside the hall, another three thousand worshippers

stomped in the cold winds off the river, pressed against the gates. The sheer numbers of bodies compelled the management to pull down the inside wall panels and open up the doors, not in compassion, but to protect the property from the pressure of bodies against the structure.

A condolence wire from President Nixon was read to a now noisy, stressful crowd of eight thousand: "The spirit of Mahalia Jackson will stir in the soul of our country the resolute will to press forward in achieving the true meaning of brotherhood to which she gave such a poignant and dynamic voice."

Civil rights leader Bayard Rustin rose, "It's a very sad time for many of us being here at this time in light of our dear friend, Mahalia Jackson's death."

Harry Belafonte, another old friend: "She was the single most powerful black woman in the United States, the woman-power for the grass roots. There was not a single field-hand, a single black worker, a single black intellectual who did not respond to her Civil Rights message."

As the service concluded, Mahalia's family stood up, visibly shaken from the attenuated service and the stress of holding together. They moved behind the wheeled coffin, down the central aisle of the giant hall, and the crowd swayed and shook as the coffin rolled past. Family, ministers, celebrated guests moved out of the hall, onto the highway; the cortège was joined by the police motorcycle escort again and a parade of vehicles and thousands of anonymous mourners in the line of march.

Thus began the twelve-mile drive to the Providence Memorial Park in Metairie, a New Orleans suburb. The police led the way through a cordon of Mardi Gras bleachers under construction for the pre-Lenten Carnival that was to begin the next day. Marching bands played funeral dirges, and from the crowd came a chant: "The angels are beginning to move over for Mahalia."

At the gravesite, the ceremony was short, but faces showed the effect of the long, exhausting, astounding day of grief and strain.

The procession reformed at the grave, moved back to the cars, and then began the joy. Bands burst forth with medleys of gospel music. The crowd took its cue, high-stepping and gyrating in the road, and the cars began their return trip to the city, to the stomp and swing of Dixieland. And as the milling people moved away from the open cemetery gates, they acted as if they knew for certain that Mahalia had taken her first steps on her long journey up the glory road.

Beautiful Spoils

"Money changes people," Mahalia agreed, at first, philosoph-
ically, sitting up high in her double condominium overlook-
ing Lake Michigan.

"You're livin' too high. You're like Zacharias. You *must*
come down," Brother John chided. He was worried. She
looked worn out. Mahalia had just made a record (it was to be
her last), and he was critical of her and it. The great Mahalia
was trying to do the modern jazz gospel. "You're tryin' to go
with rock an' roll. What's wrong with you? You're destroying
Mahalia Jackson."

"Tend to your own business," she responded. "I'm no more the Mahalia you knew as a child."

"That one was full of Grace!" he had dared to challenge.

And on the cover of the album she looked like a dead woman, according to John, what with the kind of wig she was wearing, the dress, and so on. "That woman has no life!" he had challenged, feeling that it was all inappropriate. (Ironically, it would come to pass she would be laid out in the same wig and dress.)

"Don't put bad luck on me, man! I know what Zacharias did!"

Both of them, intimate with every word, nuance, encouragement, or warning in the Bible, and accustomed to using its voices as if they were at the kitchen table with them, knew what they were sparring about: Someone who becomes so in his self, belligerent to people around him who had helped him—there comes the time when God says: *You must come down!*

John had even dared to say to her: "Remember, if you throw a brick and hide your hand, your days are numbered."

Her life had become real estate, her fortune, accountants and her new lawyer at the condominium every day. As for singing for Grace, her fees had skyrocketed so that the churches complained to her agent, but to no avail. "I know those people," she'd say. "They have *it*, they can find *it*, and if they can't, they won't have Mahalia Jackson!"

She wrestled with the two Mahalias she had become: The powerful public one with fits of anger, ruthless, unthinking; the other, lonely in the condominium, hours on the phone with her second husband, even though they were divorced; and when John was there to help her maintain the apartments and cook soul food for her, she complained of nightmares, would come looking for him at night, way in the far side of

the huge apartment, crawl into his bed, and they would talk
about the good old days: his boyhood, the beginnings—when
he slept between her and Ike—the days of music and struggle
for recognition. The talk was far away from God's warning to
Zacharias or the throwing of bricks and hiding one's hand.

Some twenty-five beneficiaries, mostly relatives (but John received nothing) shared in specific bequests in her will dated
July 24, 1971. Nine people were named beneficiaries of a
residuary trust. The Greater Salem Baptist Church was given
five thousand dollars and title to three South Side properties.
It has been variously estimated that her estate could have been
as much as nine million dollars from real estate in New Orleans, Chicago, and Los Angeles, unpaid recording royalties,
and her share of the Mahalia Jackson Chicken System set up
by Benjamin Hooks. A surprise beneficiary was Mildred
Falls, whom she hadn't talked to in a number of years—
$2000. But the record suggests that Mildred (as may have been
true with other family beneficiaries) never received the money
before she died in 1973 in a nursing home.

Undoubtedly, Mahalia had also left a large sum in cash. It
had always been important for the resolute, "back'a town"
New Orleans girl to have cash-in-hand. She never forgot her
Uncle Porter's admonition: "Don't trust anyone in this world
but yourself, and whatever work you do, remember—get
paid in cash, in full!"

Twenty years after my first meeting with Mahalia—June
1975—the second film of my projected trio called "Kinfolks"
finally became a reality. It came about when one day I received a call from an official of CBS News. "Hey," he said,
"we've been thinking over here that maybe it's time you made
that documentary on the life and music of Mahalia Jackson.

You've been pitching the project at us for twenty years. What do you say?"

You say "Yes!" when they finally come around, is what you say. So, in partnership with CBS, the network's extensive archives were made available to me. I rummaged around in the glorious jazz collections at Tulane University and the New Orleans Jazz Museum as well. I had to make the historic stills of the city of New Orleans serve in lieu of non-existing film: the turn-of-the-century street scenes of the time of Mahalia's childhood: the docks, railways, Storyville, the old brothel quarter; and the great black music-makers Mahalia had referred to as having made "indecent music" when she was a girl.

I hired a film crew and shot film of typical New Orleans street funerals. Later, I located early footage of Mahalia performing in New York's Harlem churches, where she, with Mildred Falls at the piano and organ, had made exceptional "Joyful noise unto the Lord."

It took almost a year to put the archival scenes and the new footage together. Studs Terkel became my magnanimous chronicler, the narrator of the resultant film that would be called "Got to Tell It." The film ultimately reflected our sensitive remembrance of her: "Her art was her work. Her work was her art."

It was ironic that it had to take me all those tough and stressful years of the civil rights embattlement, together with Mahalia's meteoric rise, to finally find not only a celebrated and heavyweight sponsor like CBS but also a nationwide audience now ready to respond to the black gospel and its prime interpreter. And one last thought on the subject: *Where were you when I needed you?* The greatest voice in the art had to die before there was funding available to produce such a film.

What pleasure she would have had to be a part of the making, to see the curve of the images of her life, from "back'a town" to the final scene of the picture.

It was Mahalia singing "The Lord's Prayer," recorded in a singular evening of the Newport Jazz Festival in the 1960s. The occasion marked the seventieth birthday of her friend Louis Armstrong. Earlier in the evening, a New Orleans trumpet choir had saluted him with a mélange of his favorite songs. They were followed by the virtuoso performance of the Eureka Brass Band, made up of the men who had marched and played in street bands when Louis and they were young together. Mahalia came out on stage in a pink organza dress. She and Louis linked arms and sang two duets: "Just a Closer Walk With Thee" and "Precious Lord, Walk With Me."

By then, it was raining heavily on an audience of five thousand. Pandemonium. People rose from their seats, wet to the skin, and prepared to escape to their cars and buses. To calm the crowd, Louis belted out a medley of his hit numbers, and Mahalia, not to be outdone, launched into a matchless rendition of "Hello, Dolly!"—such an atypical, surprising turn for a woman who had kept to her strict rule of singing nothing but gospel and spirituals. Rain or no rain, everyone was riveted and there was no frightening stampede to their cars.

Mahalia's feet had become so painfully swollen that she limped backstage where Louis gallantly helped her change her shoes. But he proposed she go back out there and use her influence with "the man," have Him halt the rain that continued to pelt the audience.

The two returned to the stage, to thunderous applause, and they sang an incredible duet of "When the Saints Go Marching In."

Suddenly, miraculously, the rain stopped. The crowd roared its appreciation and sat down, soaked and wanting more.

Mahalia then closed the evening with her magisterial solo: "The Lord's Prayer" in a contralto as stirring, radiant, and evoking of human emotion, dignity and beauty, as any great diva who had ever commanded and graced a stage.

Appendix

RECORDINGS FOR APOLLO RECORDS

Title	Date Recorded
I Want to Rest	10/3/46
He Knows My Heart	"
Wait Till My Change Comes	"
I'm Going to Tell God	"
What Could I Do	9/12/47
Move On Up a Little Higher (Part 1)	"
Move On Up a Little Higher (Part 2)	"
Even Me	"

Title	*Date Recorded*
I Have a Friend	"
Dig a Little Deeper in God's Love	12/10/47
Tired	"
If You See My Saviour	"
In My Home over There	"
There's Not a Friend Like Jesus	"
Amazing Grace	"
Since the Fire Started Burning in My Soul	"
I Can Put My Trust in Jesus	7/15/49
Let the Power of the Holy Ghost	"
A Child of the King	"
Get Away Jordan	"
Walk With Me	"
Prayer Changes Things	"
Shall I Meet You over Yonder	1/12/50
The Last Mile of the Way	"
Just over the Hill (Part 1)	"
Just over the Hill (Part 2)	"
I Do Don't You	"
God Answers Prayers	"
I'm Glad Salvation Is Free	"
Do You Know Him	"
I'm Getting Nearer My Home	9/11/50
I Gave Up Everything to Follow Him	"
It Pays to Serve Jesus	"
These Are They	"
He's the One	"
I Walked into the Garden	10/17/50
Bless This House	"
Go Tell It on the Mountain	"
Silent Night	"
The Lord's Prayer	"

Title	*Date Recorded*
How I Got Over	7/17/51
Just As I Am	"
Jesus Stepped Right In	"
I Bow on My Knees	"
City Called Heaven	"
It's No Secret	"
My Eye Is on the Sparrow	"
God Spoke to Me	3/21/52
In the Upper Room (Part 1)	"
In the Upper Room (Part 2)	"
Said He Would	"
He's My Light	"
If You Just Keep Still	"
I Believe	8/8/53
Beautiful Tomorrow	"
Consider Me	"
What Then	"
Hands of God	10/9/53
It's Real	"
No Matter How You Pray	"
Down to the River	5/5/53
One Day	"
My Cathedral	10/9/53
Walkin' to Jerusalem	10/12/53
I Wonder If I Will Ever Rest	"
Come to Jesus	"
I'm on My Way	6/10/54
My Story	"
Run All the Way	"
Nobody Knows	"

RECORDINGS FOR COLUMBIA

Bless This House. Columbia Records CL 899, CS 8761

"Let the Church Roll On," "God Knows the Reason Why," "Standing Here Wondering Which Way to Go," "By His Word," "Trouble of the World," "Bless This House," "It Don't Cost Very Much," "Summertime," "Sometimes I Feel Like a Motherless Child," "Just a Little While to Stay Here," "Take My Hand, Precious Lord," "Down by the Riverside," "The Lord's Prayer"

Come On, Children, Let's Sing. Columbia CL 1428, CS 8225

"Come On, Children, Let's Sing," "If We Never Needed the Lord Before," "Because His Name Is Jesus," "You Must Be Born Again," "Brown Baby," "The Christian's Testimony," "Keep a-Movin'," "A Christian Duty," "One Step," "God Is So Good"

Great Gettin' Up Morning. Columbia CL 1343, CS 8153

"Great Gettin' Up Morning," "How Great Thou Art," "I Found the Answer," "To Me It's Wonderful," "His," "God Put a Rainbow in the Sky," "He Must Have Known," "When I've Done My Best," "Just to Behold His Face," "My Journey to the Sky," "Tell the World About This"

Great Songs of Love and Faith. Columbia CL 1824, CS 8624

"Danny Boy," "The Green Leaves of Summer," "I've Done My Work," "The Rosary," "Crying in the Chapel," "A Perfect Day," "Because," "Whither Thou Goest," "Trees," "My Friend," "The House I Live In"

Newport 1958. Columbia CL 1244, CS 8071

"An Evening Prayer," "I'm on My Way," "A City Called Heaven," "It Don't Cost Very Much," "Walk over God's Heaven," "The Lord's Prayer," "Didn't It Rain," "My God Is Real," "He's Got the Whole World in His Hands," "I'm Going to Live the Life I Sing About in My Song," "Joshua Fit the Battle of Jericho," "His Eye Is on the Sparrow"

The Power and the Glory. Columbia CL 1473, CS 8264; tapes: CQ 326, RCQ 7

"Onward, Christian Soldiers," "The Holy City," "Holy, Holy, Holy," "In the Garden," "Just As I Am," "Rock of Ages," "Lift Up Your Heads," "My Country, 'Tis of Thee," "The Lord Is My Light," "Jesus, Saviour, Pilot Me," "Nearer, My God, to Thee," "Abide with Me"

Recorded in Europe During Her Last Concert Tour. Columbia CL 1726, CS 8526

"Tell the World About This," "There Is a Balm in Gilead," "Down by the Riverside," "In My Home over There," "He's Right on Time," "Eliah Rock," "It Don't Cost Very Much," "You'll Never Walk Alone," "How I Got Over"

Silent Night—Songs for Christmas. Columbia CL 1903, CS 8703

"Sweet Little Jesus Boy," "A Star Stood Still," "Hark, the Herald Angels Sing," "Christmas Comes to Us All Once a Year," "Joy to the World," "O Come, All Ye Faithful," "O Little Town of Bethlehem," "What Can I Give," "Go Tell It on the Mountain," "Silent Night, Holy Night"

Sweet Little Jesus Boy. CL 702

"Silent Night, Holy Night," "No Room at the Inn," "O Little Town of Bethlehem," "The Holy Babe," "Joy to the World," "O Come, All Ye Faithful," "Go Tell It on the Mountain," "White Christmas," "I Wonder As I Wander," "Sweet Little Jesus Boy"

You'll Never Walk Alone. Columbia CL 2552

"Trouble in My Way," "Down by the Riverside," "You're Not Living in Vain," "Without a Song," "Joshua Fit the Battle of Jericho," "You'll Never Walk Alone"

I Believe. Columbia CL 1549, CS 8349

"Trouble," "I Believe," "I'm Grateful," "I See God," "Holding My Saviour's Hand," "Somebody Bigger Than You and I," "I Asked the Lord," "I Hear Angels," "Always Look Up, Never Look Down"

Let's Pray Together. Columbia CL 2130, CS 8930

"Altar of Peace," "One God," "Let's Pray Together," "Without a Song," "Take God by the Hand," "Guardian Angels," "We Shall Overcome," "Song for My Brother," "Deep River," "No Night There," "If I Can Help Somebody"

Mahalia Sings. Columbia CL 2452, CS 9252

"Rusty Bells," "Like the Breeze Blows," "Somewhere Listening," "Shall I Become a Castaway," "Jesus Is the Light," "I Thought of You and Said a Little Prayer," "Sunrise, Sunset,"

"Just a Closer Walk with Thee," "He Is Here," "God Speaks," "This Old Building," "The Velvet Rose"

Mahalia Jackson—The World's Greatest Gospel Singer and the Falls-Jones Ensemble. Columbia CL 644, CS 8759

"I'm Going to Live the Life I Sing About in My Song," "When I Wake Up in Glory," "Jesus Met the Woman at the Well," "Oh Lord, It Is I," "I Will Move On Up a Little Higher," "When the Saints Go Marching In," "Jesus," "Out of the Depths," Walk over God's Heaven," "Keep Your Hand on the Plow," "Didn't It Rain"

Mahalia Jackson's Greatest Hits. Columbia CL 2004, CS 8804

"Walk in Jerusalem," "The Upper Room," "He Calmed the Ocean," "It Is No Secret," "How I Got Over," "Then the Answer Came," "Just over the Hill," "That's What He's Done for Me," "Move On Up a Little Higher," "Nobody Knows the Trouble I've Seen"

Make a Joyful Noise unto the Lord. Columbia CL 1936, CS 8736

"Sign of the Judgment," "That's All Right," "He Is Beside Me," "In Times Like These," "I Couldn't Keep It to Myself," "It's in My Heart," "No Other Help I Know," "It Took a Miracle," "Without God I Could Do Nothing," "Speak, Lord Jesus," "Lord, Don't Let Me Fail"

How I Got Over. Columbia KC 34073 (1976)

"How I Got Over," "I Been 'Buked and I Been Scorned," "Come on Children, Let's Sing," "Out of the Depths,"

"Move On Up a Little Higher," "His Eye Is on the Sparrow," "How Many Times," "Silent Night," "When They Ring the Golden Bells" (Producers: Jules Schwerin/Tony Heilbut)

Mahalia Jackson: Gospels, Spirituals & Hymns. Columbia Records (Digital)(1991)

Thirty-six recordings

RECORDING FOR CAEDMON RECORDS (1973)

The Life I Sing About (New Orleans reminiscences with Jules Schwerin)

"I'm Gonna Live the Life I Sing About in My Song," "Didn't It Rain," "God Put a Rainbow in the Sky," "The Lord's Prayer"

CASSETTE FOR SMITHSONIAN/
FOLKWAYS RECORDINGS
(FTS 31101 & 31102) (1976, 1992)

Got To Tell It, produced by Jules Schwerin

"He's Got the While World in His Hands," "Joshua Fit the Battle of Jericho," "His Eye Is on the Sparrow," "The Lord's Prayer" TEXT: by Mahalia Jackson (her New Orleans origins)

FILM (1976)

"Got To Tell It." CBS News.
Biography of Mahalia Jackson, with narration by Studs Terkel. (Jules Schwerin, producer, director, writer)
Distributer: Phoenix Films, New York

GRAMMY AWARD (1976)

Best Soul/Gospel Performance: "How I Got Over." National Arts and Sciences, Columbia Records. (Producers: Jules Schwerin/Tony Heilbut)

PROGRAMS FOR CBS RADIO:
"MAHALIA JACKSON SHOW" (30 MIN.)

Program #1, September 26, 1954. "I Believe," "Didn't It Rain," "You'll Never Walk Alone," "The Lord's Prayer"

Program #2, October 3, 1954. "Trees," "Joshua Fit the Battle of Jericho," "My Friend," "Sometimes I Feel Like a Motherless Child," "Summertime"

Program #3, October 10, 1954. "Jesus Met the Woman at the Well," "The Man Upstairs," "Because," "His Eye Is on the Sparrow"

Program #4, October 17, 1954. "On My Way," "Danny Boy," "Nobody Knows the Trouble I've Seen," "Brahms' Lullabye"

Program #5, October 24, 1954. "Summertime," "Down by the River," "Ezekial Saw the Wheel," "Somebody Bigger Than You and I"

Program #6, October 31, 1954. "Lucky Old Sun," "If We Never Needed the Lord Before," "The Rosary," "Lonesome Road"

Program #7, November 7, 1954. "Juanita," "I'm Gonna Live the Life I Sing About," "Just a Closer Walk with Thee," "Without a Song"

Program #8, November 14, 1954. "In the Great Getting-Up Morning," "How Are Things in Glocca Mora," "When They Ring the Golden Bells," "The End of a Perfect Day"

Program #9, November 21, 1954 (Thanksgiving Show). "My Faith Looks Up to Thee," "When the Saints Go Marching In," "Bless This House"

Program #10, November 28, 1954. "Crying in the Chapel," "It's Me Oh Lord," "City Called Heaven," "Battle Hymn of the Republic"

Program #11, December 5, 1954 (New York recordings). "Walk All Over God's Heaven," "Didn't It Rain," "One God," "Jesus Met the Woman at the Well," "Move On Up a Little Higher," "Walk All Over God's Heaven"

Program #12, December 12, 1954. "Joy to the World," "One for the Little Baby," "Go Tell It on the Mountain," "Silent Night"

Program #13, December 19, 1954. "White Christmas," "No Room at the Inn," "Oh Come All Ye Faithful," "Silent Night"

Program #14, December 26, 1954 (Request Program). "I Believe," "When the Saints Go Marching In," "Summertime," "I'm Gonna Live the Life I Sing About," "The Lord's Prayer"

Program #15, January 2, 1955. "Because," "How Many Times," "Going Home," "Hold On," "The House I Live In"

Program #16, January 9, 1955 (Columbia Records Program). "Rusty Old Halo," "Walk All Over God's Heaven," "Treasure of Love," "Woman at the Well," "Rusty Old Halo" (Repeat), "You'll Never Walk Alone"

Program #17, January 16, 1955. "Deep River," "Walk All Over God's Heaven," "I'm Grateful," "Down by the Riverside," "Go Down Moses"

Program #18, January 23, 1955 (10 min.). "Beautiful Tomorrow," "Rusty Old Halo," "God Spoke to Me One Day"

Program #19, January 30, 1955 (10 min.). "Joshua Fit the Battle of Jericho," "Walk All Over God's Heaven," "One Step"

Program #20, February 6, 1955 (10 min.). "Every Time I Feel the Spirit," "Jesus Met the Woman at the Well," "My Task"

Bibliography

Cash, W. J., *The Mind of the South*. New York: Vintage, 1941.

Garland, Phyl., *The Sound of Soul*. New York: Pocket Books, 1969.

Goreau, Laurraine, *Just Mahalia, Baby*. Gretna, La.: Pelican, 1975.

Hammond, John, with Irving Townsend, *John Hammond on Record*. New York: Summit Books, 1977.

History of Violence in America. New York: New York Times/Bantam Books, 1969.

Jackson, Mahalia, with Evan McLeod Wylie, *Movin' On Up*. New York: Hawthorn Books, 1966.

Jacobs, Paul, and Saul Landau with Eve Pell, *To Serve the Devil* (*Natives and Slaves*). New York: Random House, 1971.

Lemann, Nicholas, *The Promised Land.* New York: Alfred A. Knopf, 1991.

Lewis, Anthony, *Portrait of a Decade.* New York: Bantam Books, 1964.

Lincoln, Eric, and Lawrence H. Mamiya, *The Black Church in the African American Experience.* Durham, N.C.: Duke University Press, 1990.

Lorell, John Jr., *Black Song.* New York: Paragon House, 1986.

Right On!: Anthology of Black Literature. Edited by Bradford Chambers and Rebecca Moon. New York: New American Library, 1970.

Scheim, David E., *Contract on America.* Silver Spring, Md.: Argyle Press, 1983.